SpringerBriefs in Applied Sciences and Technology

Thermal Engineering and Applied Science

Series Editor

Francis A. Kulacki, University of Minnesota, USA

For further volumes:
http://www.springer.com/series/10305

Tamanna Alam · Poh Seng Lee
Li-Wen Jin

Flow Boiling in Microgap Channels

Experiment, Visualization and Analysis

 Springer

Tamanna Alam
Poh Seng Lee
Department of Mechanical Engineering
National University of Singapore
Singapore

Li-Wen Jin
Department of Building Environment
  and Energy Applications
Xi'an Jiaotong University
Xi'an, Shaanxi
People's Republic of China

ISSN 2193-2530   ISSN 2193-2549 (electronic)
ISBN 978-1-4614-7189-9   ISBN 978-1-4614-7190-5 (eBook)
DOI 10.1007/978-1-4614-7190-5
Springer New York Heidelberg Dordrecht London

Library of Congress Control Number: 2013944779

Printed on acid-free paper

Springer is part of Springer Science+Business Media (www.springer.com)

Preface

Thermal management has become a critical issue for high performance electronic devices. These devices have higher packaging densities and power consumptions due to the better functionalities and more compactness. Therefore, advanced thermal management techniques are necessary to ensure safe and reliable operation. Thermal management system should be compact, lightweight, and reliable for practical deployment. The use of flow boiling in microgap is a very attractive approach to satisfy these requirements. Flow boiling in microgap offers several advantages, such as extremely high heat transfer rate in compact space with a smaller coolant flow rate, greater temperature uniformity across the heat sink (both spanwise and streamwise), lower flow boiling instabilities, and ability to mitigate hotspots on the heat sink. Other advantages of this technology are the ease of fabrication and implementation (direct cooling) as fluid (dielectric coolant) can flow on the back surface of an active electronic component which in turn eliminates the need for attaching a separate heat sink and thereby removes the interface thermal resistance which is often the bottleneck for effective heat transfer/cooling.

The objectives of this brief on *Flow Boiling in Microgap Channels* are to (1) obtain better fundamental understanding of the flow boiling processes, (2) evaluate the effect of microgap dimension on the performance of microgap heat sinks, (3) evaluate the effect of surface roughness on the performance of microgap heat sinks, (4) compare performance between microgap and conventional microchannel heat sinks, (5) evaluate the microgap heat sink for instabilities and hotspot mitigation.

This brief presents an up-to-date summary of details of the confined to unconfined flow boiling transition criteria, flow boiling heat transfer and pressure drop characteristics, instability characteristics, two-phase flow pattern and flow regime map, and the parametric study of microgap dimension. Advantages of flow boiling in microgaps over microchannels are also highlighted.

Acknowledgments

The Authors gratefully acknowledge the Asian Office of Aerospace Research and Development (AOARD) and the Science and Engineering Research Council (SERC) of the Agency for Science Technology and Research (A-STAR) Singapore for their financial support for this work.

The Editorial Assistance of the staff at Springer is also gratefully appreciated.

Contents

Chapter 1
Introduction

Keywords Thermal management · Microchannel · Microgap channel · Flow boiling · Flow visualization · Heat transfer · Pressure drop · Instability · Hotspot

1.1 Background

Thermal management has become a critical issue for high performance technology in defense electronic systems designs [1]. Defense electronic devices, such as the High Energy Laser (HEL) and the Active radar system, have increased in power consumption and reduced in physical size. This increased power density has led to an increasing intensity in heat generation. Thermal management is a series of systematic steps to remove this excessive generated heat during normal operation. Heat dissipation from defense applications is at heat fluxes of the order of 1,000 W/cm^2 [2, 3]. This high heat dissipation rate is primarily due to the greater functionalities and higher packaging densities. To ensure safe and reliable operations of electronic devices, there is a need for high capacity thermal management techniques. It therefore seems likely that new techniques will be needed in the near future.

One possible technique of thermal management is to use a boiling fluid as a coolant, as this would transfer significantly more heat than its single-phase equivalent. This has led to an abundance of research into flow boiling in microchannels. Utilizing the latent heat of a coolant, boiling fluid can dissipate significantly higher heat fluxes while requiring smaller rates of coolant flow than its single-phase counterpart. Another advantage of this technique is the greater temperature uniformity across the microchannel heat sinks as the phase change of boiling fluid takes place at the fluid saturation temperature. In spite of these appealing attributes, relatively little is known about the complex nature of convective boiling and two-phase flow in microchannels and this has impeded their wide implementation in practical applications. The major issues that require immediate attention include flow instabilities and flow reversal; lack of

T. Alam et al., *Flow Boiling in Microgap Channels*,
SpringerBriefs in Thermal Engineering and Applied Science,
DOI: 10.1007/978-1-4614-7190-5_1, © The Author(s) 2014

fundamental understanding of the underlying mechanism for two-phase flow and heat transfer in the microscale domain, and the associated lack of generally accepted models for predicting two-phase pressure drop and boiling heat transfer in micro/mini channels. To address these issues, a new technique of thermal management known as "flow boiling microgap channel cooler" was recently proposed to directly cool the heat sources. Microgap channel coolers provide direct contacts among chemically inert, dielectric fluids and the back surface of an active electronic component. Thus, microgap eliminates the significant interface thermal resistance associated with Thermal Interface Materials and/or solid–solid contact between the component and a microchannel cold plate [4]. In addition, the two-phase flow boiling microgap channel coolers have potential to mitigate the flow instabilities and flow reversal issues inherent with two-phase microchannel heat sinks as the vapor generated has room to expand both spanwise and down-stream instead of being forced upstream. Moreover, it can be used for mitigating hotspots as it maintains a uniform fluid layer over the heated surface.

In recent years, a number of studies have attempted to better understand the flow boiling mechanism in microgap [4–9]. Kim et al. [4, 6] experimentally investigated the two-phase thermo-fluid characteristics of a dielectric liquid, FC-72, flowing in an asymmetrically heated chip scale microgap with channel heights varying from 110 to 500 μm. This exploratory study revealed that the intermittent and annular flow regimes dominate the 110 to 500 μm channel behavior for the two-phase flow of FC-72, with a liquid volumetric flow rates from 0.17 to 0.83 ml/s.

Bar-Cohen and Rahim [5] performed a detailed analysis of microchannel/microgap heat transfer data for two-phase flow of refrigerants and dielectric liquids, gathered from the open literature and sorted by the Taitel and Dukler flow regime mapping methodology. They showed that the annular flow regime is the dominant regime for this thermal transport configuration and its prevalence is seen to grow with decreasing channel diameter and to become dominant for refrigerant flow in channels below 0.1 mm diameter. Sheehan and Bar-Cohen [7] investigated a 210 micron microgap channel, operated with mass flux 195.2 kg/m^{2}s and heat flux varying from 10.3 to 26 W/cm^2 using infrared thermography to observe wall temperature fluctuation and locate nascent dryout regions. They concluded from their results that wall temperature fluctuations vary independently with both thermodynamic quality of the flow and the wall heat flux and these fluctuations reflect a complex interplay of channel and local instabilities with periodic local dryout and re-wetting.

Kabov et al. [8] studied a detailed map of the flow sub-regimes in a shear-driven liquid film flow of water and FC-72 obtained for a 2 mm channel operating at room temperature. They showed that shear-driven films are more suitable for cooling applications than falling liquid films. Utaka et al. [9] experimentally investigated flow boiling of water in narrow gaps of 0.5, 0.3, and 0.15 mm and measured the thickness of the micro-layer by application of the laser extinction method. They showed that the initial micro-layer thickness decreases with the decreasing gap size and the heat transfer was enhanced due to the micro-layer evaporation. Although progress has been made to characterize the heat transfer and pressure drop during flow boiling in microgaps, a fundamental understanding of boiling mechanisms along with the flow visualization for such microgaps are unavailable.

Recently, Bar-Cohen and Rahim [10] proposed microgap cooler to control the temperature of on-chip hotspots on the back of the chip. To assess the efficacy of micro-gap cooler for thermal management of hotspots, they simulate the thermal performance of a notional advanced semiconductor chip cooled by the micro-gap coolers. They found that micro-gap coolers, along with effective thermal spreading in the chip, appears to offer the potential for successfully limiting the chip and hotspot temperature rise to acceptable levels for a wide range of operating conditions. More experimental investigation is necessary to evaluate the possibility and efficacy of microgap heat sink for thermal management of hotspots.

As illustrated in the above literature review, the current understanding of two-phase heat transfer in microgap is still far from being well-established. Crucial issues, such as the flow boiling behaviors in microgap based on high speed flow visualization, dimensional and surface roughness parameter influence in microgap, are still unclear. Thus, the concept of the two-phase microgap channel technique is still very new, and more investigation should be carried out to advance the fundamental understanding of the underlying mechanisms. These investigations could consider systematic experimental investigation to characterize flow boiling phenomenon in microgap channel as well as flow visualization investigation for flow regime detection. These fundamental investigations may further lead to continuous improvement of thermal management in defense electronic systems.

1.2 Objectives

The specific objectives of this brief are to:

1. Present the experimentally investigated local flow boiling heat transfer and pressure drop characteristics in microgap channel over a range of gap dimensions and flow rates to achieve better fundamental understanding.
2. Correlate the high speed visualized flow patterns and their transitions in microgap channel with simultaneous heat transfer and pressure drop measurements.
3. Investigate the effect of microgap size on flow boiling characteristics and optimize the microgap size for better heat transfer characteristics.
4. Examine the effect of surface finish on flow boiling heat transfer characteristics in microgap channel.
5. Examine the effectiveness of two-phase microgap channel cooling technology for hotspot remediation and thermal management.

Assess the ability of the two-phase microgap coolers in mitigating flow instabilities and flow reversal.

References

1. Lee J, Mudawar I (2009) Low-temperature two-phase microchannel cooling for high-heat-flux. Thermal management of defense electronics. IEEE Trans Compon Packag Technol 32(2):453–465
2. Pokharna H, Masahiro K, DiStefanio E, Mongia R, Crowley BJ, Chen W, Izenson M (2004) Microchannel cooling in computing platforms: performance needs and challenges in implementation. Second international conference on microchannels and minichannels (ICMM), pp 109–118
3. Kandlikar SG, Bapat AV (2007) Evaluation of jet impingement, spray and microchannel chip cooling options for high heat flux removal. Heat Transf Eng 28(11):911–923
4. Kim DW, Rahim E, Bar-Cohen A, Han B (2008) Thermofluid characteristics of two-phase flow in micro-gap channels. 11th IEEE intersociety conference on thermal and thermomechanical phenomena in electronic systems (I-THERM), pp 979–992
5. Bar-Cohen A, Rahim E (2009) Modeling and prediction of two-phase refrigerant flow regimes and heat transfer characteristics in microgap channel. Heat Transfer Eng 30(8):601–625
6. Kim DW, Rahim E, Bar-Cohen A, Han B (2010) Direct submount cooling of high-power LEDs. IEEE Trans Compon Packag Technol 33(4):698–712
7. Sheehan J, Bar-Cohen A (2010) Spatial and temporal wall temperature fluctuations in two-phase flow in microgap coolers. In: Proceedings of the ASME 2010 international mechanical engineering congress and exposition (IMECE), pp 12–18
8. Kabov OA, Zaitsev DV, Cheverda VV, Bar-Cohen A (2011) Evaporation and flow dynamics of thin, shear-driven liquid films in microgap channels. Exp Thermal Fluid Sci 35:825–831
9. Utaka Y, Okuda S, Tasaki Y (2009) Configuration of the micro-layer and characteristics of heat transfer in a narrow gap mini/micro-channel boiling system. Int J Heat Mass Transf 52:2205–2214
10. Bar-Cohen A, Wang P (2012) Thermal management of on-chip hot spot. J Heat Transf 134:1–11

Chapter 2
Design and Operating Parameters

Keywords Flow loop · Microchannel test section · Microgap channel test section · Test surface · Design parameters · Operating parameters

The experiment facility, test section, design, and operating parameters are briefly described in this chapter. More details of test procedure, calibration procedure and heat loss calculation are available in Alam et al. [1, 2].

2.1 Experimental Flow Loop

The closed experimental flow loop consists of a reservoir, a gear pump, an inline 15 μm filter, a liquid flow sensor, a pre-heater, test section, and liquid-to-air heat exchanger. Deionized water in the reservoir is fully degassed before initiating each experimental run. The schematic diagram of flow loop is shown in Fig. 2.1. Temperature measurements are done before the entry and after the exit of the heat exchanger, at the inlet and outlet of the test section using type-T thermocouples. The inlet and outlet pressure are measured using Pressure transmitter. Constant-voltage power is supplied to the 25 integrated heaters on the backside of the chip to provide the desired heat flux for the flow boiling experiment. An integrated 5 × 5 diode temperature sensor array is used to measure the temperature distribution on the chip. A high-speed camera is mounted over microgap test piece to capture visual data at frame rate 5,000 fps. The data from all different sensors are collected using a computer-based Data acquisition and measurement control system.

2.2 Test Section

Schematic diagram of the microgap and microchannel test sections are presented in Fig. 2.2. A silicon test piece, a top cover and a base plate are the main three components of test section. The test piece consists of 12.7 mm × 12.7 mm silicon

T. Alam et al., *Flow Boiling in Microgap Channels*,
SpringerBriefs in Thermal Engineering and Applied Science,
DOI: 10.1007/978-1-4614-7190-5_2, © The Author(s) 2014

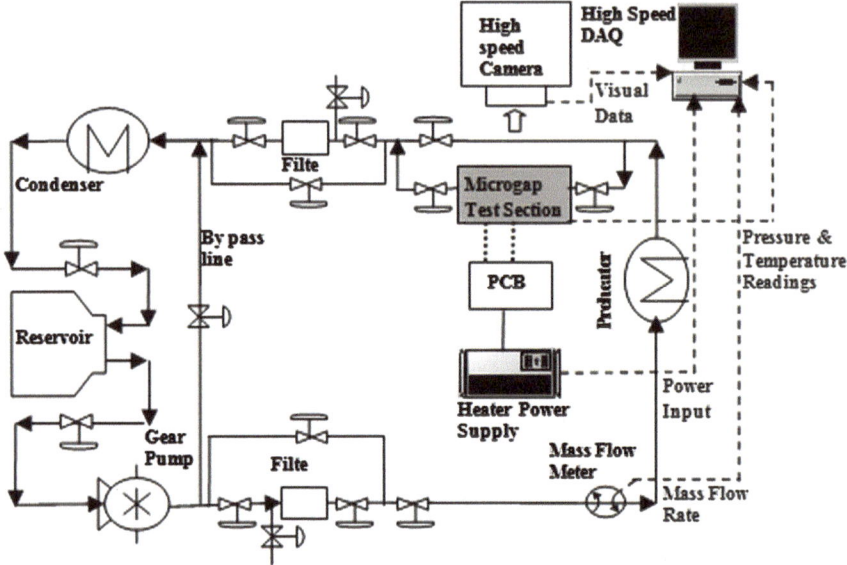

Fig. 2.1 Schematic diagram of the flow loop

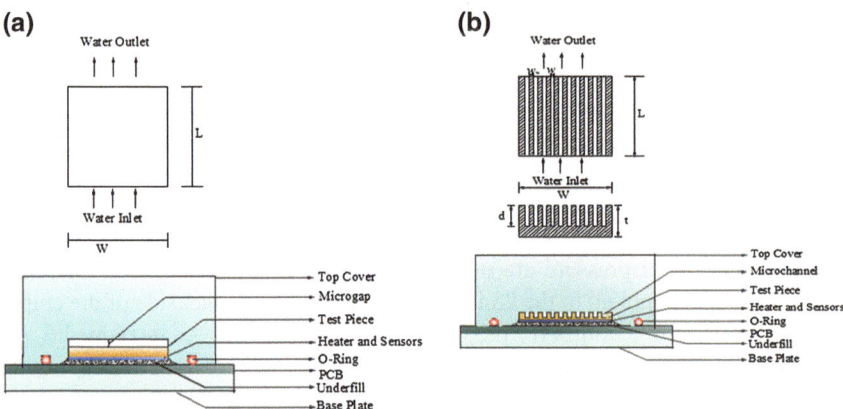

Fig. 2.2 Schematic diagram of **a** microgap channel test section, **b** microchannel test section

heat sink mounted on a printed circuit board (PCB). The thermal test dies are fabricated using a five-inch type-P silicon wafer with orientation 111. The dies are 625 μm thick and are diced in an array of 5 × 5. The diced chips were mounted on PCB using 63Sn/37Pb solder bumps. The test piece includes 25 heat sources and temperature-sensing diodes as shown in Fig. 2.3. Each of the 25 heater/tempera-ture sensor elements is 2.54 mm x 2.54 mm in size and incorporates a heating element and integrated diode sensors for on-die temperature monitoring. Top cover is made of polycarbonate and a 1.27 cm x 1.27 cm Pyrex glass of thickness

Fig. 2.3 Test piece with
5 x 5 array of heating
elements and integrated diode
temperature sensors

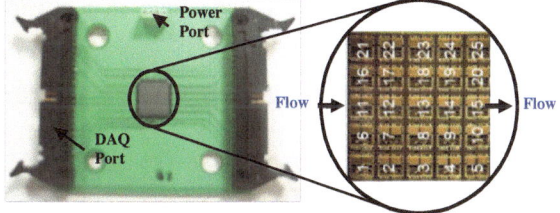

0.3 cm is fitted in it to get clear flow visualization. A desired dimension groove is maintained at the bottom of the Pyrex glass. Top cover including Pyrex glass is positioned over PCB including silicon test piece in such a manner that it makes the desired microgap dimension over the test surface. During experiment, water flow between the diced silicon chip (opposite surface to the attached heater and sensor) and bottom of the Pyrex glass. For microchannel test piece, channels are cut into the top surface of the chip. A square O-ring is used to seal the test piece between the cover and the chip to ensure good sealing of the heat sink. Four set screws are used at the four corners to bolt the top cover with base plate through test piece. After assembling the test section, the actual dimension is measured using 3-axis Measuring Microscope with camera. Inlet and outlet manifolds are formed within top cover across the heat sink. Moreover, holes are drilled into the top cover for locating the pressure taps, thermocouples, fluid inlet, and fluid outlet.

Heat loss characterization and Diode temperature sensors calibration of the test section have been performed before initiating experiments. All the heat transfer results presented in this work are based on the diode position 15, which is the location, last downstream along the center row (as shown in Fig. 2.3) in the heat sink as it is most likely to experience the highest degree of saturated boiling as shown in Fig. 2.4a. However, the wall temperature variation of test sections in the lateral direction is determined to be only within ≈ 0.4 °C. Wall temperature

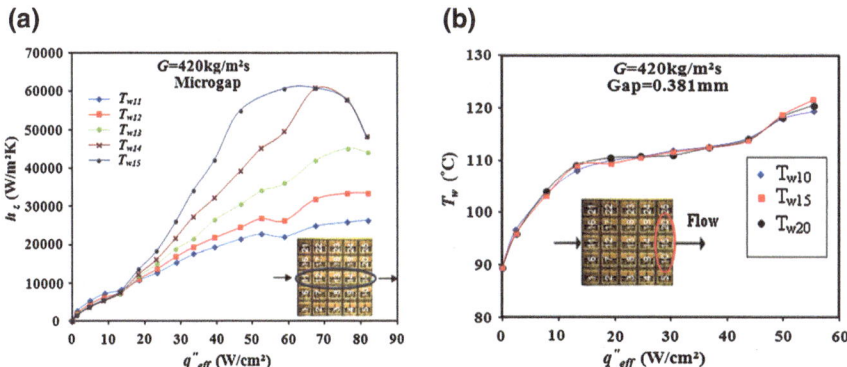

Fig. 2.4 a Variation of heat transfer coefficient curves along the centre row of streamwise direction at mass flux, $G = 420$ kg/m^{2}s for microgap heat sink, **b** Variation of local wall temperature at three spanwise locations

variation with effective heat flux at three spanwise locations with the diode sensors 10, 15, and 20 are shown in Fig. 2.4b.

2.3 Microgap Test Surface

The original silicon surface of surface roughness, $R_a = 0.6$ µm is modified to $R_a = 1.0$ and 1.6 µm to examine the effect of surface finish. The original bare silicon surface is roughened with 180 and 100 grit sandpaper to obtain surface roughness, $R_a = 1.0$ and 1.6 µm respectively. The magnified images of surface finish and quantitative 3D images of the three different rough chip surfaces are shown in Fig. 2.5.

2.4 Design and Operating Parameters

See Tables 2.1, 2.2, 2.3, 2.4, 2.5.

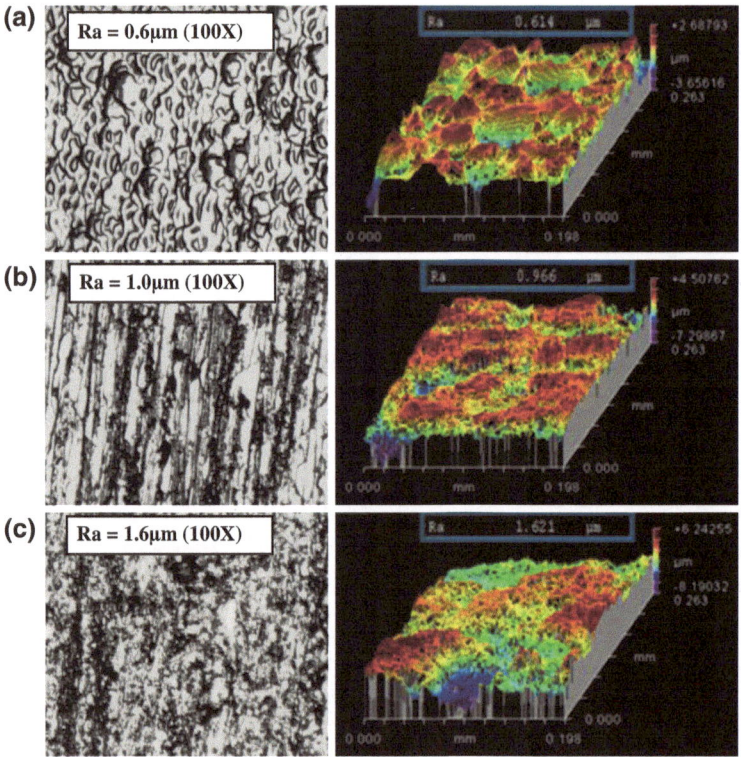

Fig. 2.5 Magnified images of chip surfaces with oblique plots (3D), **a** bare silicon, $R_a = 0.6$ µm, **b** scratched silicon, $R_a = 1.0$ µm, **c** scratched silicon, $R_a = 1.6$ µm

Table 2.1 Microgap and microchannel dimensions and experimental conditions used for flow boiling comparative study

Test piece	Case	L (mm)	W (mm)	D (μm)	w (μm)	w_w (μm)	d (μm)	N	G (Kg/ m²s)	$T_{f,i}$(°C)	q''_{eff} (W/ cm²)
Microgap	1.1	12.7	12.7	190.34	–	–	–	–	420	86.49	0–85
	1.2								690	86.73	0–55
	1.3								970	86.58	0–55
Microchannel	2.1	12.7	12.7	–	208.28	215.72	385.70	30	420	86.40	0–45
	2.2								690	86.06	0–55
	2.3								970	86.63	0–65

Table 2.2 Microgap dimensions and experimental conditions used for flow boiling heat transfer and pressure drop study

Case	Length, L (cm)	Width, W (cm)	Gap depth, D (μm)	Mass Flux, G (kg/m²s)	Inlet fluid temp, $T_{f,i}$ (°C)	q''_{eff} (W/ cm²)
1.1	1.27	1.27	190.34	420	86.49	0–90
1.2				690	86.73	0–65
1.3				970	86.58	0–65
2.1	1.27	1.27	285.22	420	86. 90	0–100
2.2				690	86.56	0–100
2.3				970	86.34	0–100
3.1	1.27	1.27	380.79	420	86.75	0–60
3.2				690	86.79	0–90
3.3				970	86.42	0–110

Table 2.3 Microgap dimensions and experimental conditions used for optimizing microgap channel dimension and operating condition study

Case	Length, L (cm)	Width, W (cm)	Depth, D (μm) (actual values)	Mass flux, G (kg/m²s)	Inlet fluid temp, $T_{f,i}$ (°C)	q''_{eff} (W/ cm²)
1	1.27	1.27	80 (79.1)	390,655,903	90.4	0–28
2	1.27	1.27	100 (96.7)	390	90.4	0–28
3	1.27	1.27	150 (152.3)	384,652,905	90.9	0–28
4	1.27	1.27	200 (196.2)	390,651,902	90.8	0–47
5	1.27	1.27	300 (298.9)	385,657,902	90.9	0–59
6	1.27	1.27	360 (358.6)	387,650,905	90.6	0–59
7	1.27	1.27	500 (501.9)	382,654,901	90.4	0–59
8	1.27	1.27	700 (715.2)	384,651	90.5	0–65
9	1.27	1.27	850 (866.5)	386	91.1	0–71
10	1.27	1.27	1,000 (1,001.9)	384	91.2	0–71

Table 2.4 Microgap dimensions and experimental conditions used for surface roughness effect on flow boiling heat transfer and pressure drop study

Length, L (cm)	Width, W (cm)	Depth, D (μm) (actual values)	R_a (μm) (actual values)	R_t (μm)	Mass flux, G (kg/m²s)	$T_{f,i}$ (°C)	q''_{eff} (W/cm²)
1.27	1.27	200 (196.2)	0.6 (0.5723)	4.8953	390,651	90.8	0–47, 0–50
		200 (201.1)	1.0 (1.0058)	9.0463	391,655	90.1	0–50, 0–50
		200 (204.8)	1.6 (1.6554)	16.491	390,652	90.1	0–45, 0–50
		300 (298.9)	0.6 (0.5723)	4.8953	385,657	90.9	0–59, 0–71
1.27	1.27	300 (302.8)	1.0 (1.0058)	9.0463	384,655	90.6	0–65, 0–71
		300 (305.1)	1.6 (1.6554)	16.491	381,660	90.4	0–72, 0–71
		500 (501.9)	0.6 (0.5723)	4.8953	382,654	90.4	0–59, 0–71
1.27	1.27	500 (490.3)	1.0 (1.0058)	9.0463	395, 661	90.1	0–59, 0–83
		500 (504.9)	1.6 (1.6554)	16.491	396, 660	90.5	0–59, 0–83

Table 2.5 Test piece dimensions and experimental conditions used for hotspots mitigation study

Test piece	L (mm)	W (mm)	D (μm)	w (μm)	w_w (μm)	d (μm)	N	G (Kg/m^{2}s)	$T_{f,i}$ (°C)
Microgap	12.7	12.7	190.34	–	–	–	–	400–1,000	86
	1.27	1.27	285.22						
	1.27	1.27	380.79						
Microchannel	12.7	12.7	–	208.28	215.72	385.70	30	400–1,000	86

References

1. Alam T, Lee PS, Yap CR, Jin LW (2012) Experimental investigation of local flow boiling heat transfer and pressure drop characteristics in microgap channel. Int J Multiph Flow 42:164–174
2. Alam T, Lee PS, Yap CR, Jin LW (2013) A comparative study of flow boiling heat transfer and pressure drop characteristics in microgap and microchannel heat sink and an evaluation of microgap heat sink for hotspot mitigation. Int J Heat Mass Transf 58:335–347

Chapter 3
Characteristics of Two-Phase Flow Boiling in Microgap Channel

Keywords Microgap channel · Flow boiling · Flow visualization · Confinement effect · Heat transfer · Pressure drop

The current chapter presents the heat transfer and pressure data of three different dimension microgap heat sinks collected during the experimental program. Experimental results are discussed to characterize the two phase flow boiling heat transfer and pressure drop behaviors in microgap channel. Vapor confinement criteria was adopted from Harirchian and Garimella [1]. and experimental data was plotted based on this criterion to predict the vapor confinement inside the microgap channel under different operating conditions and microgap dimension. The heat transfer mechanisms are explained based on the vapor confined criteria and flow boiling regime. High speed visualizations are shown to validate the explanation.

3.1 Vapor Confinement Criterion in Microgap Channel

Vapor bubbles grow from nucleation sites at the heated surface. As the heat flux is increased, a confined slug/annular flow appear in these microgaps. During confined flow, vapor flow in the center of the microgap, creating a vapor core and separated from microgap wall by a thin liquid layer. Thin film evaporation occurs throughout the liquid vapor interface of the bubbles. Heat transfer rate increases significantly when the thin liquid layer is the only major thermal resistance to conduction. If vapor bubble is relatively smaller than the microgap depth, unconfined flow occurs. Criterions are developed by many researchers to define vapor confinement. For example, Monde et al. [2] have used Bond number, Bo to relate gap depth to bubble departure diameter. Several researchers including Monde et al. [2], Geisler and Bar-Cohen [3] identify Bo $\approx$ 1 as the transition criteria between confined and unconfined flow. In this experiment, Bond number is below 1, for all gap size meets the confinement criterion.

Harirchian and Garimella [1] have shown that flow confinement depends not only on the dimension and fluid properties but also on the mass flux since the

T. Alam et al., *Flow Boiling in Microgap Channels*,
SpringerBriefs in Thermal Engineering and Applied Science,
DOI: 10.1007/978-1-4614-7190-5_3, © The Author(s) 2014

Fig. 3.1 Vapor confinement
criterion in microgap channel

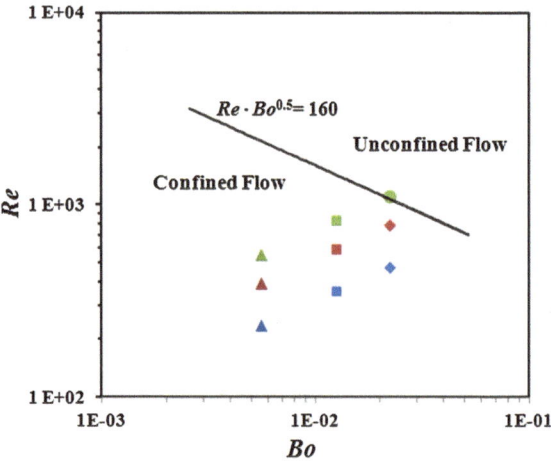

bubble diameter varies with flow rate. Based on their study, they proposed a new correlation, represented by,

$$\text{Bo}^{0.5} \cdot Re = \frac{1}{\mu_f} \left[\frac{g\left(\rho_f - \rho_g\right)}{\sigma} \right]^{0.5} GD^2 = 160$$

This new flow boiling transition criteria recommends that for $\text{Bo}^{0.5} \cdot Re < 160$, vapor bubbles are confined in microgap and microchannel. By adopting this criterion, all the experimental data have been plotted and found all microgap channels with confined flow as shown in Fig. 3.1.

3.2 Boiling Curve

The effect of microgap sizes on boiling curve for mass flux of 420 kg/m²s at inlet fluid temperature 86 °C is shown in Fig. 3.2. The point in the figure where the wall temperature exhibits a sudden change in slope from its single-phase dependence can be identified as the onset of nucleate boiling (ONB). After the ONB, the wall temperature increase gradually with the increase of heat flux for all gap size and wall temperature varies for different gap sizes. This dependence of the wall temperature on the heat flux for these microgap sizes at low mass flux 420 kg/m²s may indicate that convective boiling rather than nucleate boiling is the main heat transfer mechanisms. It can be seen from the figures that wall temperature remain more uniform and low as the gap size decrease. This observation can be attributed to rapid increase of vapor quality and early transition to annular flow in smaller gap with a fixed mass flux.

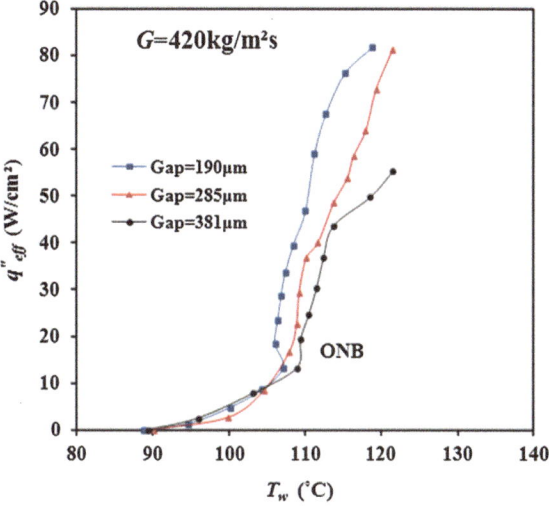

Fig. 3.2 Effect of microgap sizes on local wall temperatures at $G = 420$ kg/m^{2}s

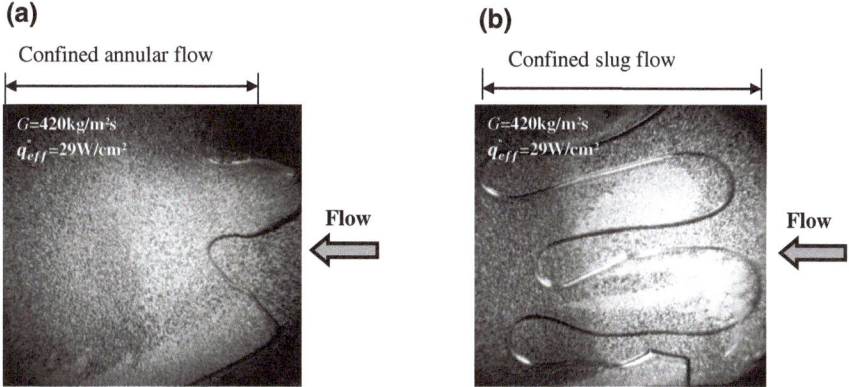

Fig. 3.3 Flow visualization at microgap of depth **a** 190 μm, **b** 285 μm

Flow visualizations have been performed as part of this work to support and confirm this observation. Flow visualization images at microgap of depth 190 and 285 μm for a fixed heat flux, $q''_{eff}=29$ W/cm^2 are shown in Fig. 3.3. It can be seen from figure that for a fixed heat flux condition, smaller gap experiences a confined annular flow whereas larger gap experiences confined slug flow. During confined annular flow, it is obvious that the void fraction, and thus the vapor quality are much larger than in confined slug flow. Moreover, due to confined annular flow, thin film evaporation occurs throughout the liquid–vapor interface of the vapor core and is the most effective heat transfer mechanism occurring in the boiling

Fig. 3.4 Effect of mass fluxes on local wall temperatures at microgap of depth 381 μm

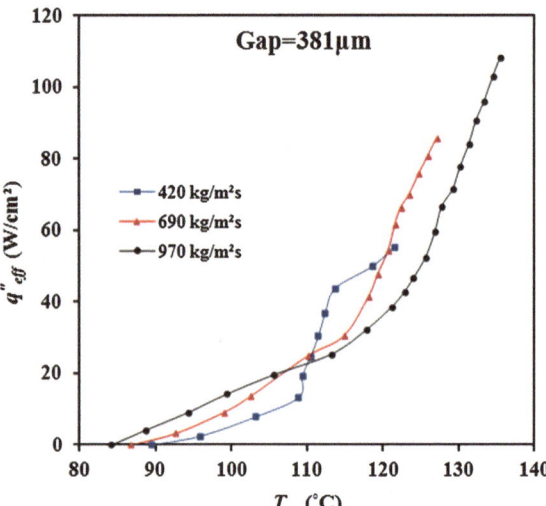

process. This increases the heat transfer performance of smaller gap and maintain lower wall temperature compare to higher gap at same mass and heat flux.

Boiling curves are plotted for 381 μm gap test section at three different mass fluxes as shown in Fig. 3.4. As can be seen from the figure, a higher heat flux can be achieved with the increase of mass flux in the single phase region and ONB occurs earlier at lower mass flux. After the ONB, boiling curves show sensitivity with mass flux and lower mass flux maintain lower wall temperature for a fixed heat flux. This behavior is attributed to earlier transition to annular flow as well as rapid increase of vapor quality for lower mass flux at a fixed heat flux.

3.3 Two-Phase Local Heat Transfer Coefficient

The effect of microgap sizes on local heat transfer coefficient for a fixed mass flux 420 kg/m²s at inlet fluid temperature 86 °C is shown in Fig. 3.5. The local heat transfer coefficient is calculated at position 15, the central diode in the last downstream location as indicated in Fig. 2.3. The presence of local temperature sensors allows the local heat transfer coefficients to be computed. It can be seen from Fig. 3.5, for low heat flux input, the local heat coefficient increases almost linearly with heat flux. On the other hand, at higher heat flux input, the curve shows a change in slope after the ONB as the heat transfer coefficient increase rapidly after boiling commences and the local heat transfer coefficient is increased with decreasing the microgap size. Due to the smaller microgap size relative to the bubble diameter at departure, bubbles occupy the microgap and create confinement effects. This confinement effect gives the higher local heat transfer coefficient for smaller microgap size. In small microgap size, bubble nucleation at the walls is not

Fig. 3.5 Effect of microgap sizes on local heat transfer coefficients at $G = 420$ kg/m²s

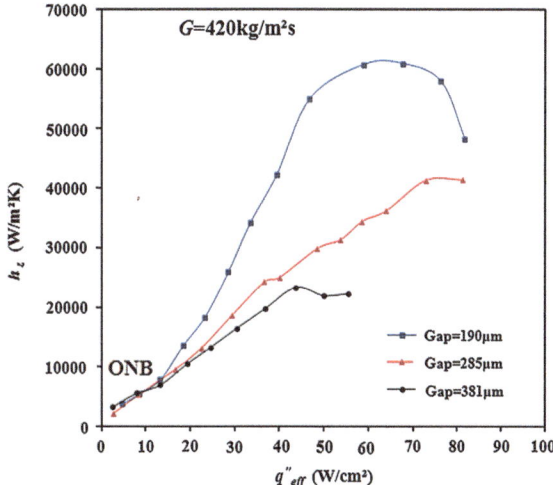

the only heat transfer mechanism; in addition the evaporation of this liquid film at the walls generates slug and annular flows and this also contributes to the heat transfer. Therefore, the value of local heat transfer coefficient is larger for this smaller depth of microgap. The heat transfer coefficient starts to decrease at around 50–60 W/cm². This is attributed to partial dryout that happens in microgap wall with higher heat flux.

The influence of mass fluxes on local heat transfer coefficient for 381 μm gap test section at inlet fluid temperature 86 °C is shown in Fig. 3.6. From figure, it can be seen that the heat transfer coefficients show dependency on mass flux after the ONB and the local heat transfer coefficients decrease with increase of mass flux.

Fig. 3.6 Effect of mass fluxes on local heat transfer coefficients at microgap of depth 381 μm

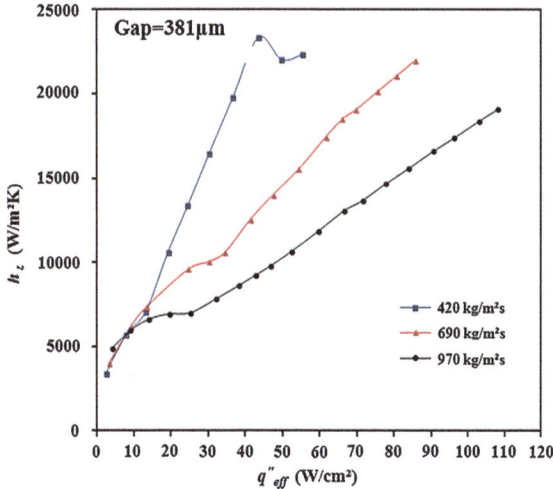

This result can be explained as- the vapor quality at the outlet is higher for a lower mass flux at a given heat flux. Because of these decreases of vapor quality, the local heat transfer coefficient decrease significantly at higher mass flux. Similar phenomenon has been observed by Bertsch et al. [4]. The heat transfer coefficient is decreased at high heat flux for mass flux 420 kg/m²s is because of early transition to annular flow and consequently early partial wall dryout. This dryout phase has not been observed for higher mass fluxes here because of limitation in the input heat flux to prevent high wall temperature that would damage the test die.

3.4 Two-Phase Pressure Drop Characteristics

Figure 3.7 illustrates the effect of microgap sizes on pressure drop for a fixed mass flux 690 kg/m²s at inlet fluid temperature 86 °C as a function of effective heat flux. In two phase region, pressure drop increases with decrease in microgap dimension at a fixed heat flux. This is due to the dominance of confined slug and annular boiling and high rate of vapor generation in smaller microgap for a fixed heat and mass flux as illustrated in Fig. 3.3. Moreover, with the increase of gap size, pressure drop curve become insensitive to heat flux. For gap 381 μm, pressure drop remain almost uniform with increasing heat flux. Similar trend has been [5] where with the increase of micochannel width, slope of the pressure drop curve is also decreased.

Figure 3.8 shows the variation of pressure drop as a function of heat flux for a range of mass fluxes for microgap size of 190 μm. It is observed that in the two phase region, the pressure drop increases with increasing heat flux as vapor content increase subsequently. It is also observed from figure that for a fixed heat flux, pressure drop increase with increase of mass flux. Since the test section has an

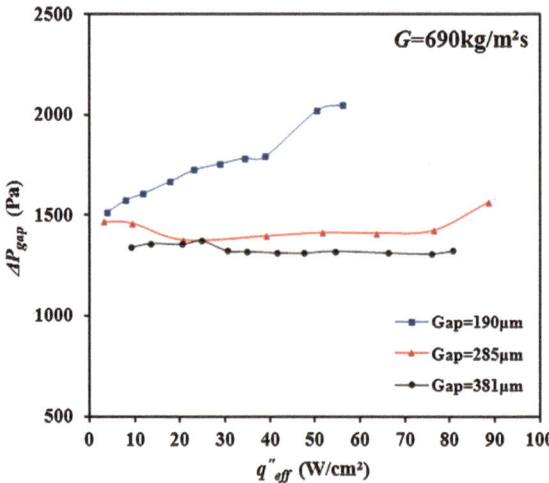

Fig. 3.7 Effect of microgap sizes on pressure drop at $G = 690$ kg/m²s

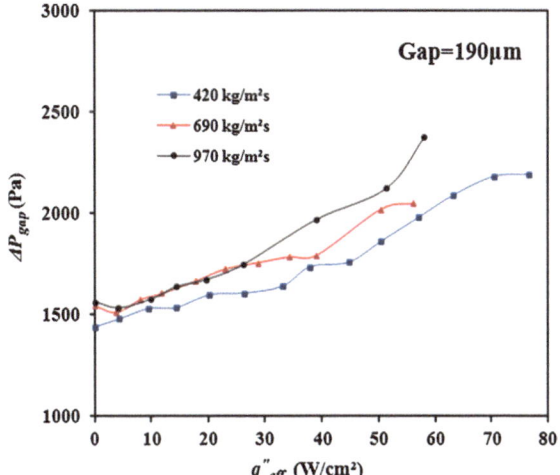

Fig. 3.8 Effect of mass fluxes on pressure drop at microgap of depth 190 μm

upstream subcooled region because of subcooled inlet fluid, subcooled region increase with increase of mass flux for same heat flux which may attributed to higher frictional pressure drop at higher mass flux.

References

1. Harirchian T, Garimella SV (2010) A comprehensive flow regime map for microchannel flow boiling with quantitative transition criteria. Int J Heat Mass Transf 53:2694–2702
2. Monde M, Kusuda H, Uehara H (1982) Critical heat flux during natural convective boiling in vertical rectangular channels submerged in saturated liquid. J Heat Transf Trans ASME 104(2):300–303
3. Geisler KJL, Bar-Cohen A (2009) Confinement effects on nucleate boiling and critical heat flux in buoyancy-driven microchannels. Int J Heat Mass Transf 52:2427–2436
4. Bertsch SS, Groll EA, Garimella SV (2009) Effects of heat flux, mass flux, vapor quality, and saturation temperature on flow boiling heat transfer in microchannels. Int J Multiph Flow 35:142–154
5. Harirchian T, Garimella SV (2008) Microchannel size effects on local flow boiling heat transfer to a dielectric fluid. Int J Heat Mass Transf 51:3724–3735

Chapter 4
Comparison of Flow Boiling Characteristics Between Microgap and Microchannel

Keywords Microgap channel · Microchannel · Flow boiling · Flow visualization · Wall temperature uniformity · Heat transfer · Pressure drop

Two-phase microgap heat sink has a large potential to minimize the drawbacks associated with two-phase microchannel heat sink, especially flow instabilities, flow reversal and lateral variation of flow and wall temperature between channels. This new concept of the two-phase microgap heat sink is very promising due to its high heat transfer rate and ease of fabrication. However, comparison of the performance of the microgap heat sink (heat transfer and pressure drop characteristics) with some conventional heat sink has not been investigated extensively.

This chapter presents the comparison of the performance of the microgap heat sink (heat transfer and pressure drop characteristics) with conventional straight microchannel heat sink. The comparison is done with the same footprint, the same inlet mass flux, and the same effective heat flux supplied based on the footprint. High speed visualizations are shown to validate the experiment results and explanation. These studies are carried out with the inlet DI water temperatures 86 °C at different mass fluxes ranging from 400 to 1,000 kg/m^{2}s, for effective heat flux 0 to 85 W/cm^2.

4.1 Flow Visualization of Boiling Processes in Microgap and Microchannel

Figure 4.1 illustrates the flow visualization of boiling processes at different imposed heat fluxes taken from the top of the microchannel at $G = 420$ kg/m^{2}s. The flow pattern at a heat flux 9 W/cm^2 in Fig. 4.1a shows that bubbles are nucleated on the channel walls. These bubbles are observed to grow and occupy almost the entire microchannel width. These bubbles are observed to coalesce and

T. Alam et al., *Flow Boiling in Microgap Channels*,
SpringerBriefs in Thermal Engineering and Applied Science,
DOI: 10.1007/978-1-4614-7190-5_4, © The Author(s) 2014

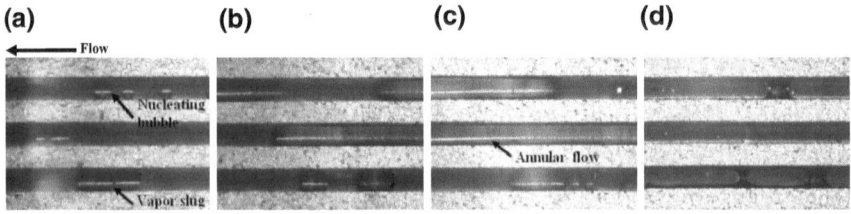

Fig. 4.1 Flow visualization of boiling process for microchannel, $G = 420$ kg/m^{2}s at heat fluxes **a** $q''_{\text{eff}} = 9$ W/cm^2, **b** $q''_{\text{eff}} = 13$ W/cm^2, **c** $q''_{\text{eff}} = 18$ W/cm^2, **d** $q''_{\text{eff}} = 23$ W/cm^2

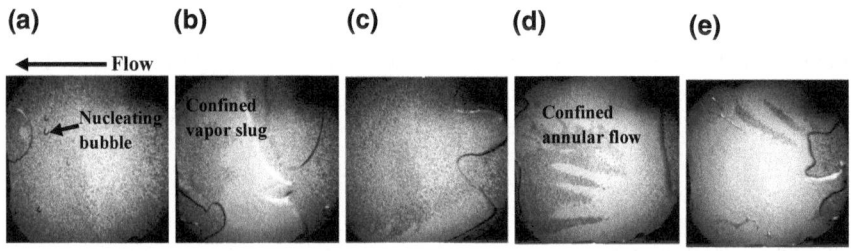

Fig. 4.2 Flow visualization of boiling process for microgap, $G = 420$ kg/m^{2}s at heat fluxes **a** $q''_{\text{eff}} = 13$ W/cm^2, **b** $q''_{\text{eff}} = 22$ W/cm^2, **c** $q''_{\text{eff}} = 29$ W/cm^2, **d** $q''_{\text{eff}} = 40$ W/cm^2, **e** $q''_{\text{eff}} = 60$ W/cm^2

vapor slugs are formed in the microchannel. At higher heat fluxes of 13, 18 and 23 W/cm^2; elongated slugs and annular flow patterns are observed in the microchannel. Based on these flow visualizations, it is noted that an early establishment of slug/annular flow occurs in this microchannel of very small diameter.

Flow visualization of boiling processes at different imposed heat fluxes taken from the top of the microgap at $G = 420$ kg/m^{2}s are shown in Fig. 4.2. At the heat flux 13 W/cm^2, discrete bubbles are observed to nucleate over the microgap wall, detach and move downstream. During sliding along the microgap heated surface, detached bubbles grow slightly as can be seen at the exit of the gap. At the heat flux 22 W/cm^2, nucleating bubbles are observed to confine in the microgap. These confined bubbles then expand and coalesce to form vapor slug. At higher heat fluxes of 29, 40 and 60 W/cm^2; confined annular flow patterns are observed to dominate the heat transfer mechanism in microgap heat sink.

4.2 Comparison of Boiling Curve Between Microgap and Microchannel

Comparison of boiling curves between the microchannel and the microgap channel heat sink at different mass fluxes are illustrated in Fig. 4.3. Tests are conducted at same inlet fluid temperature and local wall temperature is plotted against effective

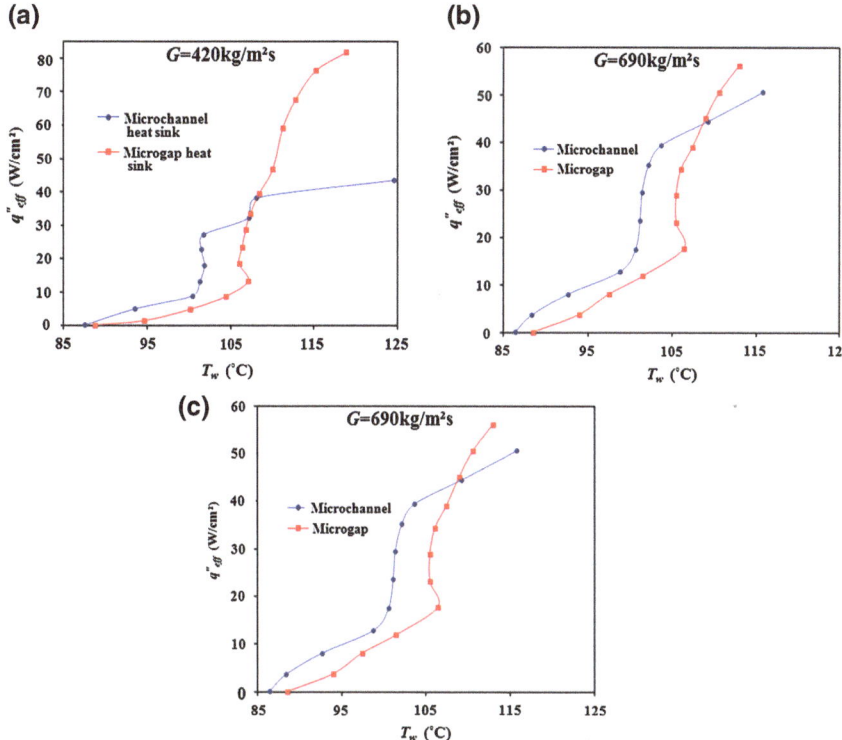

Fig. 4.3 Comparison of boiling curve between microchannel and microgap at mass flux, **a** $G = 420$ kg/m^{2}s, **b** $G = 690$ kg/m^{2}s, **c** $G = 970$ kg/m^{2}s

heat flux. The wall temperatures presented here are those measured near the exit at diode position 15. As can be seen from the figure, at the onset of nucleate boiling (ONB) the wall temperature exhibits a sudden change in slope from its single phase for both the heat sinks and lower wall temperature is needed to commence boiling over the heated surface for microchannel heat sink. It can also be seen from figure that microchannel maintains lower chip wall temperature for a fixed heat flux before encountering dryout phase. This is due to the fact that fluid gets more surface area to flow and can remove more heat from surface leading to lower surface temperature at low heat flux in microchannel test section. Moreover, early establishment of slug/annular flow and consequent rise of vapor quality in microchannel of very small diameter [1, 2] also attributes the better heat transfer performance at low heat flux. This early establishment of slug/annular flow is also observed by flow visualization presented in Fig. 4.1. However, dryout phase strikes very early for this heat sink at mass flux 420 kg/m^{2}s as shown in Fig. 4.3a. In contrast, lower heat removal rate is achieved at low heat flux region in microgap test section due to the smaller surface area. In addition, due to confined slug and annular boiling dominance in microgap heat sink at higher heat flux region as shown in Fig. 4.2, the microgap heated wall is constantly covered with confined bubbles and thin film evaporation happens throughout the liquid vapor interface. Consequently, heat removal rate increase and dryout phase delays at this heat sink at high heat flux. This results a better heat transfer performance of microgap heat sink at higher heat fluxes for a given mass flux. However, dryout phase delays in microchannel heat sink at higher heat fluxes by increasing the mass fluxes as observed in Fig. 4.3b and c.

4.3 Comparison of Two-Phase Local Heat Transfer Coefficient

Local heat transfer coefficient curves for microchannel and microgap heat sink at two different mass fluxes are presented in Fig. 4.4. The local heat transfer coefficients are calculated at position 15, the central diode in the last downstream location as indicated in Fig. 2.3. Local heat transfer coefficient is plotted against wall heat flux. It can be seen from the Fig. 4.4a that the heat transfer coefficient increases with increasing heat flux. At low to medium heat flux, microchannel gives better heat transfer coefficient as slug/annular flow commences earlier in this small diameter heat sink as shown by flow visualization in Fig. 4.1. Further increment of heat flux in this microchannel leads to a decreasing trend of local heat transfer coefficient because of partial dryout. On the contrary, heat transfer coefficient is higher for microgap heat sink than microchannel at high heat flux as microgap delays the dryout phase. In microgap, evaporation of this liquid film at the walls due to confined slug/annular flow contributes to the heat transfer at high heat flux region. Therefore, the value of local heat transfer coefficient is higher for

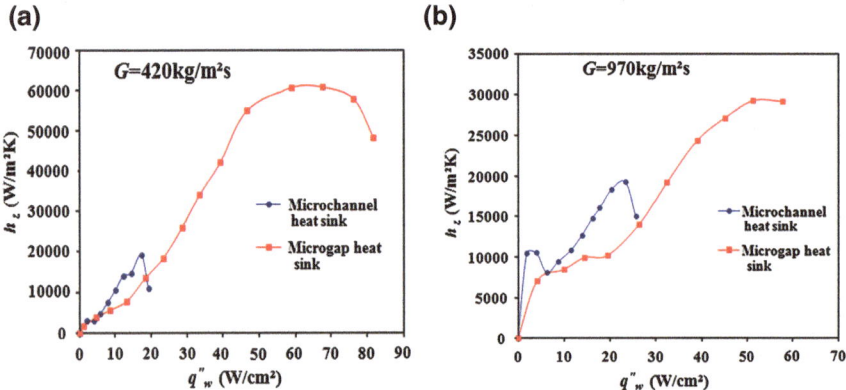

Fig. 4.4 Comparison of local heat transfer coefficient between microchannel and microgap as a function of wall heat flux at mass flux, **a** $G = 420$ kg/m^{2}s, **b** $G = 970$ kg/m^{2}s

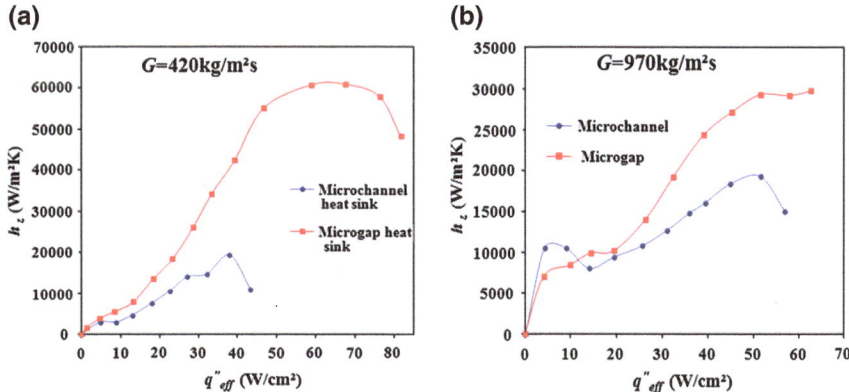

Fig. 4.5 Comparison of local heat transfer coefficient between microchannel and microgap as a function of effective heat flux at mass flux, **a** $G = 420$ kg/m²s, **b** $G = 970$ kg/m²s

this microgap than microchannel heat sink. Similar phenomenon is also observed for higher mass flux as shown in Fig. 4.4b.

The dependence of heat transfer coefficient on heat sink size should be considered in terms of a given amount of heat dissipation from the chip, i.e., a fixed value of effective heat flux, q''_{eff} from a design point of view. Thus, the heat transfer coefficients (as a function of effective heat flux) for microgap and microchannel heat sink at two different mass fluxes are presented in Fig. 4.5. It can be seen from figure that the heat transfer coefficient is higher for microgap heat sink for a given heat dissipation from the chip.

4.4 Two-Phase Pressure Drop Characteristics Comparison Between Microgap and Microchannel

Comparison of pressure drop between microchannel and microgap heat sink as a function of effective heat flux at two different mass fluxes is presented in Fig. 4.6. From figure, it can be seen that the pressure drop increases with heat flux. This is attributed to the acceleration effect of vapor and the two-phase frictional pressure drop in microchannel and microgap heat sink. Similar phenomenon has been observed by Harirchian et al. [1] and Lee and Garimella [3]. Moreover, it is observed that pressure drop is higher for microchannel than microgap heat sink at a given heat flux because of the large surface area in microchannel which induced more two-phase frictional pressure drop than microgap. In addition, the slope of the pressure drop curve is higher for microgap at high heat flux. This due to the rapid increase of vapor quality during confined annular boiling in microgap at high heat flux.

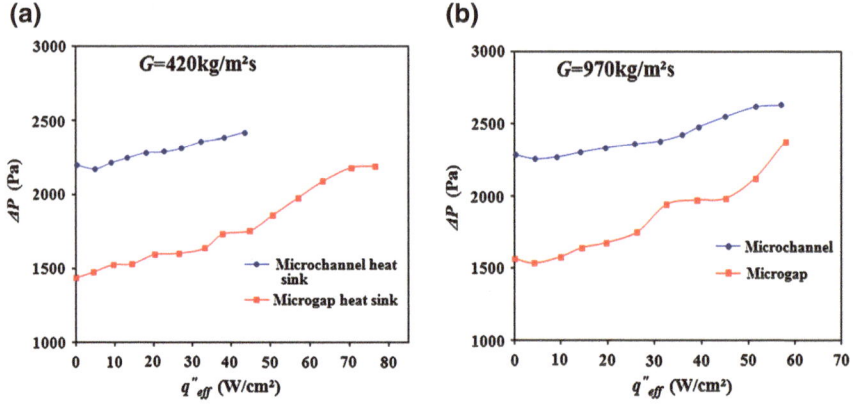

Fig. 4.6 Comparison of pressure drop between microchannel and microgap at mass flux, **a** $G = 420$ kg/m²s, **b** $G = 970$ kg/m²s

4.5 Wall Temperature Uniformity Comparison Between Microgap and Microchannel Heat Sink

Variation of local wall temperature with effective heat flux at mass flux, $G = 420$ kg/m²s for microgap and microchannel heat sink is plotted in Fig. 4.7. Wall temperatures are taken at five streamwise locations along the centre row with the diode sensors 11, 12, 13, 14, and 15. Figure 4.7a illustrates that after the ONB in microgap, wall temperatures for all the locations converge and maintain uniform wall temperature throughout the microgap wall at all heat flux condition. Confined slug and annular boiling are the two main heat transfer mechanisms in microgap heat sink. Thus, microgap heated wall is constantly covered up with confined bubbles and thin film evaporation happens throughout the liquid vapor interface. Thin liquid layers have low resistance to thermal diffusion and evaporation of liquid into the vapor core can promote the removal of substantial thermal energy from the walls. Consequently, microgap heat sink minimizes temperature gradient over the heated surface. Whereas, Fig. 4.7b illustrates that after the ONB in microchannel, all curves are converging but still there is a large variation of wall temperature along the streamwise location. This variation of wall temperature occurs due to the presence of different boiling regimes over the heated surface along the streamwise direction. Moreover, due to earlier partial dryout in microchannel at higher heat flux, the wall temperatures suddenly become very high and a large wall temperature variation is observed at these heat fluxes.

Maximum temperature reduction and minimization of temperature gradient on the heated surface of the device are the two important objectives in electronic cooling [4]. Local wall temperature variation with effective heat flux at mass flux, $G = 420$ kg/m²s for microgap and microchannel heat sink is plotted in Fig. 4.8.

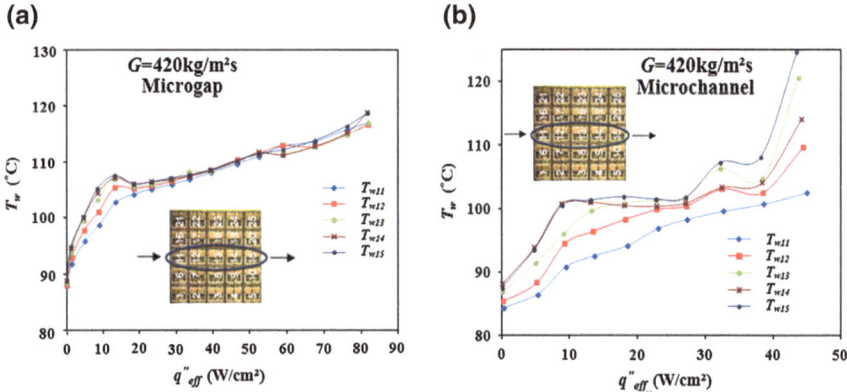

Fig. 4.7 Variation of local wall temperature along the centre row of streamwise direction at mass flux, $G = 420$ kg/m²s for **a** microgap channel heat sink, **b** microchannel heat sink

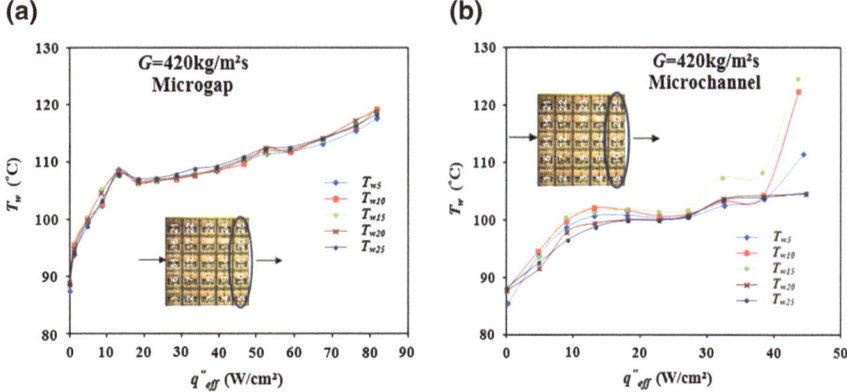

Fig. 4.8 Variation of local wall temperature along the most downstream spanwise direction at mass flux, $G = 420$ kg/m²s for **a** microgap channel heat sink, **b** microchannel heat sink

Wall temperatures are taken at five spanwise locations with the diode sensors 5, 10, 15, 20, and 25. Experimental results shown in Fig. 4.8a that microgap heat sink maintain uniform wall temperature and minimize temperature gradient along the spanwise locations at all heat flux conditions. In contrast, microchannel heat sink shows a significant lateral variation of wall temperature at high heat flux in Fig. 4.8b as dryout is approached. This is because of the presence of different boiling regime between channels in lateral direction.

References

1. Harirchian T, Garimella SV (2008) Microchannel size effects on local flow boiling heat transfer to a dielectric fluid. Int J Heat Mass Transf 51:3724–3735
2. Qu W, Mudawar I (2004) Transport phenomena in two-phase micro-channel heat sinks. J Electron Packag 126:213–224
3. Lee PS, Garimella SV (2008) Saturated flow boiling heat transfer and pressure drop in silicon microchannel arrays. Int J Heat Mass Transf 51:789–806
4. Garimella SV, Sobhan CB (2003) Transport in microchannels—a critical review. Ann Rev Heat Transf 13:1–50

Chapter 5
Optimization of Microgap Channel Dimension and Operating Condition

Keywords Microgap channel · Confinement effect · Flow boiling · Flow visualization · Wall temperature uniformity · Heat transfer · Pressure drop

The current chapter presents the heat transfer and pressure data of ten different dimension microgap heat sinks collected during the experimental program. Experimental results are presented and discussed to optimize the microgap heat sink dimension and flow condition based on two-phase flow boiling heat transfer and pressure drop characteristics in microgap channel. Vapor confinement criteria was adopted from Harirchian and Garimella [1] and experimental data was plotted based on this criterion to predict the vapor confinement inside the microgap channel under different operating conditions and microgap dimension. The heat transfer mechanisms are explained based on the vapor confined criteria and high speed visualizations are shown to validate the explanation.

These studies are carried out with the inlet DI water temperatures 91 °C at different mass fluxes ranging from 390–900 kg/m²s for ten different microgap dimensions from a range of 80–1,000 µm at imposed effective heat flux, q''_{eff} ranging from 0 to 70 W/cm².

5.1 Vapor Confinement in Microgap Channel

Vapor confinement criteria in microgap channel for mass fluxes, $G = 390$, 650, and 900 kg/m²s and for ten different microgap dimensions from a range of 80–1,000 µm are illustrated in Fig. 5.1. The confinement criterion based on dimension, fluid properties, and mass flux as described by Harirchian and Garimella [1] is used in this work. This flow boiling transition criterion recommends that for $Bo^{0.5} \cdot Re < 160$, vapor bubbles are confined in microgaps. It is

T. Alam et al., *Flow Boiling in Microgap Channels*,
SpringerBriefs in Thermal Engineering and Applied Science,
DOI: 10.1007/978-1-4614-7190-5_5, © The Author(s) 2014

Fig. 5.1 Vapor confinement
criterion in microgap channel

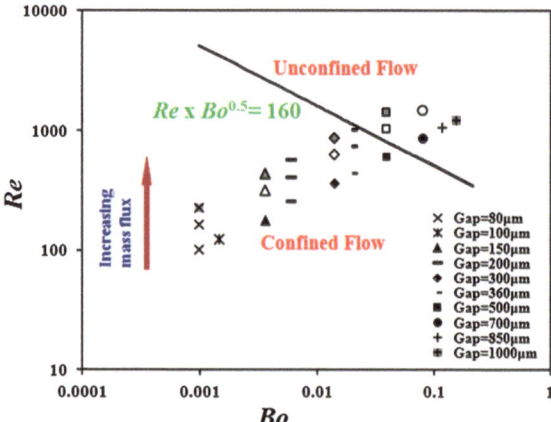

Fig. 5.2 Visualization of
boiling processes

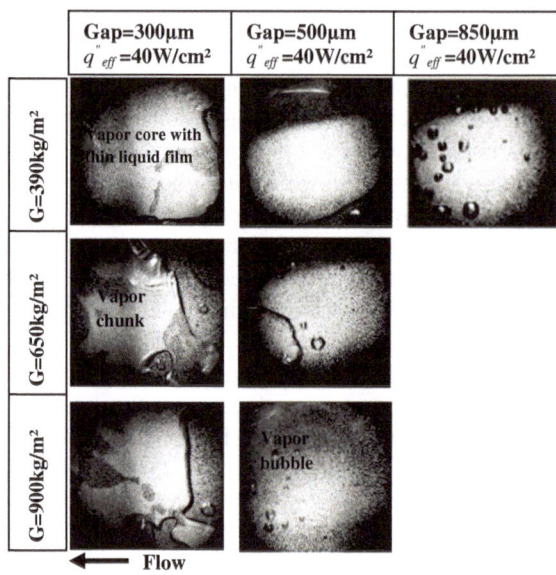

clearly noted from the figure that in microgap sizes 360 μm and below, confinement effects are shown to occur whereas microgaps with sizes 700 μm and above are shown to have unconfined behavior within the range of mass fluxes tested. However, the flow in a 500 μm microgap is shown to transform from confined to unconfined flow as mass flux increases from 390 kg/m²s to higher. These behavior patterns are also supported by the visual data presented in Fig. 5.2.

5.2 Flow Visualization of Boiling Processes in Microgap Channels

Figure 5.3 illustrates the flow visualization of boiling processes for different microgaps ranging from 80–1,000 μm at $G = 390$ kg/m^{2}s and $q''_{\text{eff}} = 28$ W/cm^2. At gap size 700 μm and above, bubbles are observed to nucleate on the microgap heated surface. These bubbles are rarely observed to coalesce with each other and also do not show any confinement inside the microgap wall. Nucleating bubbles grow up to a certain size, detach from heated surface and move downstream. At microgap size 500 μm, bubbles nucleate over the heated surface, grow, and confine inside the gap. This confined bubble then expands and moves downstream. Confined slug and churn flow shows dominance at 300 and 360 μm gap. Multiple nucleating bubbles are grown violently and confined in the gap, they expand and merge with each other to form vapor slug/chunk. This vapor slug/chunk expands further, distributes over the gap flowing with bulk fluid and moves downstream. Confined annular flow, which is characterized by thin liquid film on the microgap wall, and a vapor core in the rest of gap cross section, is observed at microgap size 200 μm and below. In these gaps, a single bubble is sufficient to create a confinement effect and confined annular flow. A bubble appears on the heated surface, grows, and expands explosively and covers up the entire area of microgap.

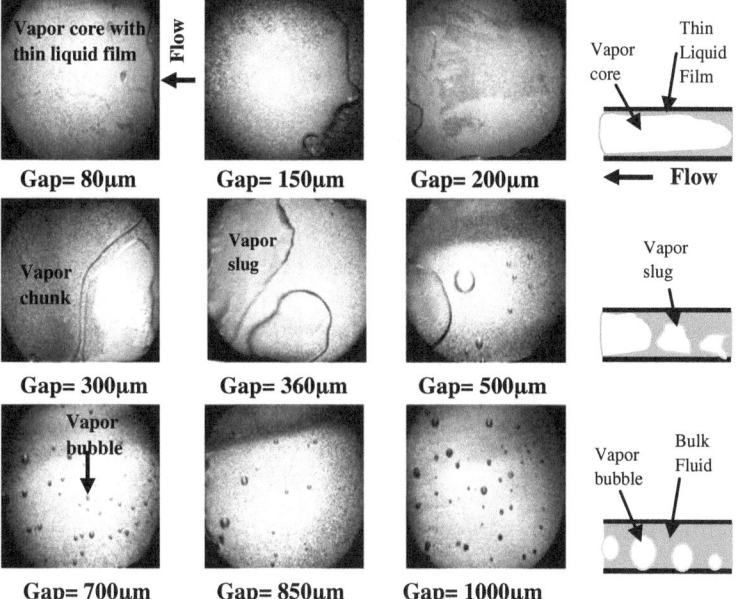

Fig. 5.3 Flow visualization of boiling process for different sizes of microgap at $G = 390$ kg/m^{2}s, $q''_{\text{eff}} = 28$ W/cm^2

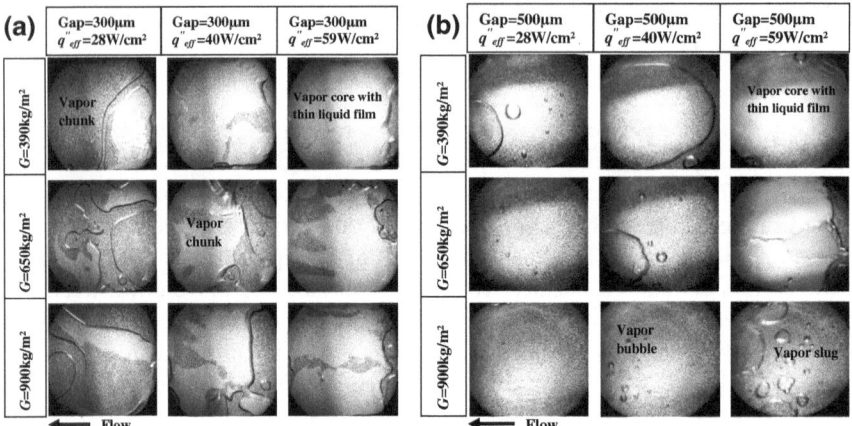

Fig. 5.4 Visualization of boiling process at various imposed heat and mass fluxes for **a** 300 μm gap, **b** 500 μm gap

Consequently, the microgap heated wall is constantly covered with confined bubbles and thin film evaporation happens throughout the liquid vapor interface. Thereafter, the confined bubble finally moves downstream from the heated wall under the effect of liquid flow and new bubbles start to repeat the process.

Visualization of boiling processes at various imposed heat flux and mass fluxes for 300 and 500 μm gap are demonstrated in Fig. 5.4a and b respectively. It is noted from the Fig. 5.4a that confined churn/annular flow are the two dominant flow boiling mechanisms for all mass fluxes and heat fluxes for microgap 300 μm. At 300 μm gap, the nucleating bubbles grow and confined in the gap at even very low heat fluxes. These confined bubbles then expand and coalesce to form vapor slug. With the further increases of heat flux, the liquid in the slugs between the bubbles begins to shrink as the expanded bubble occupies the entire microgap to develop into a confined annular flow pattern. However, the images at larger microgap, 500 μm in Fig. 5.4b indicate that flow boiling pattern changes with both the heat flux and mass flux. Bubbles nucleate over the heated surface, grow, and confine inside the 500 μm gap at only 390 kg/m²s. This confined bubble then expands and forms slug and annular pattern with the increase of heat flux. At higher mass fluxes, nucleating bubbles do not show any confinement inside the 500 μm microgap wall and forms slug/annular pattern by coalescing of bubbles with each other at high heat flux.

5.3 Flow Boiling Curves

Figure 5.5 presents the microgap size effect on boiling curve for a fixed mass flux, $G = 390$ kg/m²s. Results are obtained for nine different microgap dimensions from a range of 80–1,000 μm to identify the best suitable dimension range for

Fig. 5.5 Microgap size effect on boiling curves

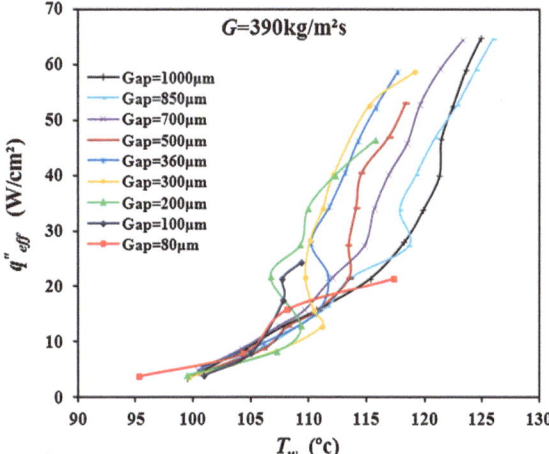

microgap. The wall temperatures presented here are those measured near the exit at diode position 15. It is noted from figure that for single phase region, the slopes of the boiling curves are fairly constant and performance increases as gap size increases. Temperature overshoot is observed at the onset of nucleate boiling (ONB) and ONB shifts towards the high heat flux as microgap size increases. It is also observed from the figure that lower wall temperature is needed to commence boiling over the heated surface for smaller gap. After the ONB, the wall temperature increases with increasing heat flux for all microgaps presented here and the boiling curve shifts significantly to the left as microgap sizes are reduced. This phenomenon indicates that the heat transfer performance is substantially better for smaller gaps in two-phase region as smaller gaps maintain lower wall temperature. Flow visualization in Fig. 5.3 revealed that this improvement of performance mainly due to early establishment of confined slug and annular flow in smaller microgaps. After a certain heat flux for each microgap size exhibit a further change in the slope of the curve and wall temperature increases very quickly with a small increase of heat flux. This is mainly results from partial wall dryout as observed by flow visualization. This partial dryout phase occurs earlier for smaller gaps for a fixed mass flux. From the above investigation and as shown in Fig. 5.3, it can be concluded that below 100 μm sized microgap is ineffective as partial dryout strikes very early and above 500 μm sized microgap, no vapor confinement is observed. The microgap size effect on boiling curve presented here is also observed for other higher mass fluxes.

Figure 5.6a and b show the effect of mass fluxes on boiling curve at microgap, 300 and 500 μm. As can be seen from the figures, more dissipation of heat flux is achieved with the increase of mass flux at a fixed wall temperature in the single phase region and ONB occurs earlier at lower mass flux. After the ONB, more heat is dissipated as mass flux decreases at a fixed wall temperature and smaller mass

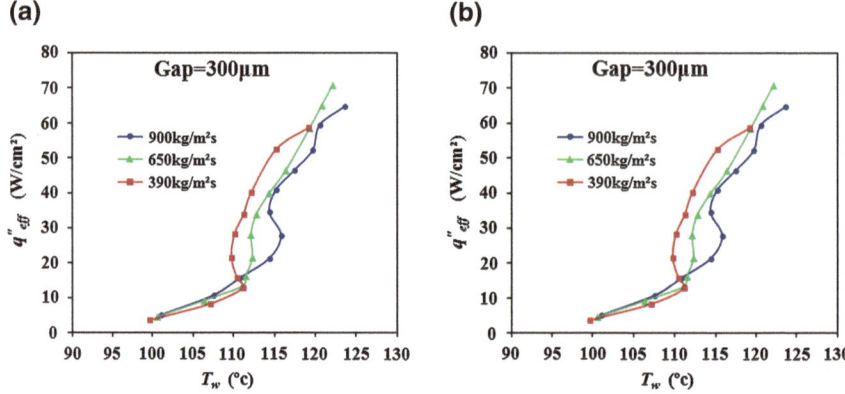

Fig. 5.6 Effect of mass flux on boiling curve at microgap, **a** 300 μm, **b** 500 μm

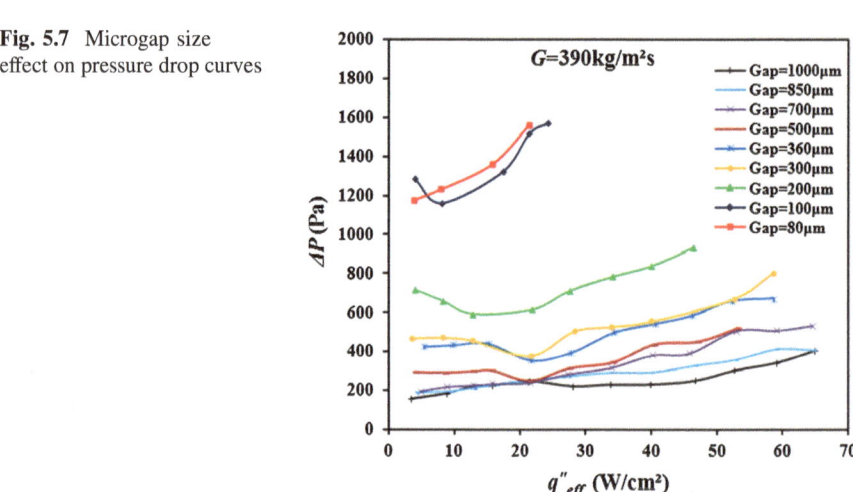

Fig. 5.7 Microgap size effect on pressure drop curves

flux maintain lower wall temperature. This may be because of higher vapor quality and earlier transition to annular flow for lower mass flux at a fixed heat flux as observed by flow visualization. Moreover, in two-phase region, the wall temperature increases gradually with the increases of heat flux for all mass flux at both the gap sizes except 900 kg/m^{2}s at 500 μm. As the boiling start in this microgap, the wall temperature has a weak dependence on the heat flux and is relatively invariant. Nucleate boiling dominance may be the reason for this weak dependence as observed by flow visualization.

5.4 Pressure Drop Curves

Microgap size effect on pressure drop across the microgap heat sink for a fixed mass flux, $G = 390$ kg/m^{2}s as a function of heat flux is shown in Fig. 5.7. Pressure drop is measured between the two manifolds upstream and downstream and loss associated with sudden contraction and expansion is corrected. It is observed from the figure that, pressure drop decreases slightly with the increases of heat flux in the single phase region due to the reduction of water viscosity with increase in temperature. In the two-phase region, pressure drop increases with heat flux due to the dominance of acceleration effect of vapor content. It is also seen from figure in both single and two-phase regions that pressure drop increases with decreasing microgap size at a fixed heat flux. The slope of the pressure drop curve also increases in two-phase region as microgap size decreases. This is due to early transition to confined slug and annular flow and rapid increase of vapor quality in smaller gap as observed by flow visualization. It is noted from the figure that below 200 μm gap, pressure drop across microgap is very high and as the gap size increases further above 200 μm; pressure drop falls and remains low for all gap sizes up to 1,000 μm.

Figure 5.8 illustrates the variation of pressure drop with heat flux for a range of mass fluxes for microgap, 300 μm. It is observed from the figure that for a fixed heat flux, pressure drop increases with increases of mass flux in two-phase region. Since the test section has an upstream subcooled region because of subcooled inlet fluid, subcooled region increase with increase of mass flux for same heat flux which may attributed to higher frictional pressure drop at higher mass flux.

Fig. 5.8 Effect of mass flux on pressure drop curves at microgap, 300 μm

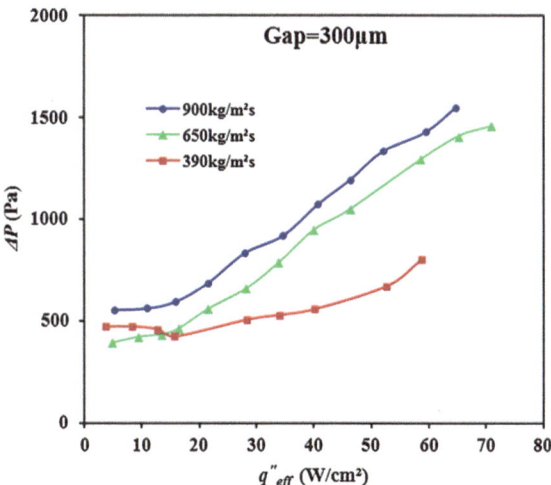

5.5 Local Heat Transfer Coefficient Curves

The effect of microgap size on local heat transfer coefficient for a fixed mass flux, $G = 390$ kg/m^{2}s as a function of heat flux is illustrated in Fig. 5.9a. The local heat transfer coefficient computed here is based on wall temperatures measured near the exit at diode position 15. It is seen from the figure that local heat transfer coefficient increases with increasing heat flux in both single and two-phase regions as expected. In single phase region, local heat transfer coefficient is independent of microgap size. In contrast, local heat transfer coefficient is highly sensitive to microgap size in two-phase region and smaller microgap gives higher heat transfer coefficient at a given heat flux. This much higher local heat transfer coefficient in smaller microgap results from the fact that the establishment of confined slug and annular flow on the heated surface occur earlier in smaller gap as observed by flow visualization. Hence thin film evaporation throughout the liquid vapor interface on the heated surface becomes dominant heat transfer mechanism in smaller gap leading to higher heat transfer coefficient. Utaka et al. [2] measured the initial microlayer thicknesses 23, 18, and 9 μm for gap sizes of 0.5, 0.3, and 0.15 mm, respectively in the constant thickness region. Thus, the micro-layer thickness is strongly affected by the gap size, and decreases with decreasing gap size [2]. Heat transfer coefficient through microlayer evaporation is modeled as follows,

$$h = \frac{k}{\delta}$$

where h is the heat transfer coefficient, k is thermal conductivity of the liquid and δ is the liquid film thickness. Therefore, in the micro-layer dominant region, the vaporization rate is increased, and higher boiling heat transfer is possible due to the thinner micro-layer [2–4]. For microgaps of size 700 μm and above, the local heat transfer coefficient is independent of gap dimension. Flow visualization revealed that these gaps do not face confinement effect and nucleate boiling is the dominant heat transfer mechanism for these gaps. A decrease in local heat transfer coefficient is detected after a certain heat flux for each microgap. This is mainly due to partial wall dryout on the heated surface. Liquid film thickness is also affected by heat flux, and decreases with increasing heat flux [2, 3]. Due to an increase in heat flux, the thinner liquid film disappears for a short time and a dryout region appears which results in the decrease of heat transfer coefficient. It is noted from figure that partial dryout appear earlier for microgap size 100 μm and below and hence ineffective.

Figure 5.9b illustrates two heat transfer coefficient curves for microgap, 300 and 1,000 μm at mass flux, $G = 390$ kg/m^{2}s. Flow visualization images taken from the top of the microgap heat sinks are illustrated with each boiling curve. While 1,000 μm gap presents the typical flow patterns (bubbly, slug\transition, and annular flows), the bubbly flow patterns for 300 μm are less evident since at even very low heat fluxes the flow is seen to have slug\transition flow characteristics.

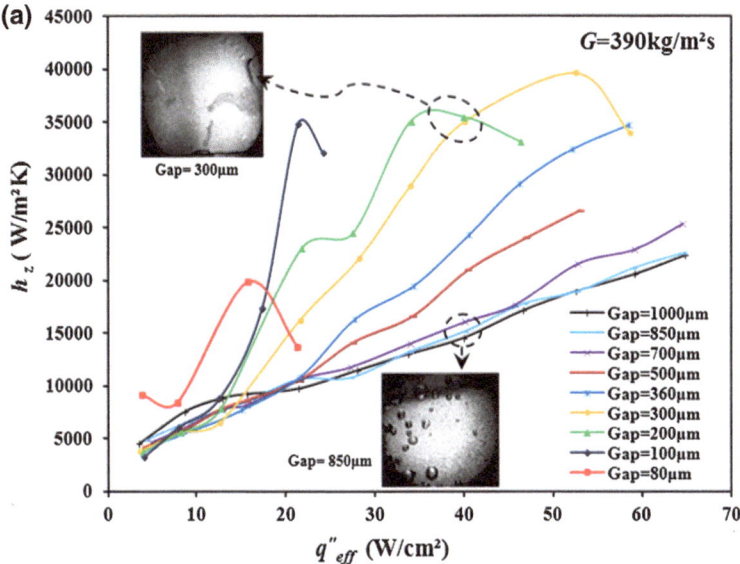

Fig. 5.9 a Microgap size effect on local heat transfer coefficient curves, **b** Heat transfer coefficient curve with flow patterns for microgap, 300 and 1,000 μm

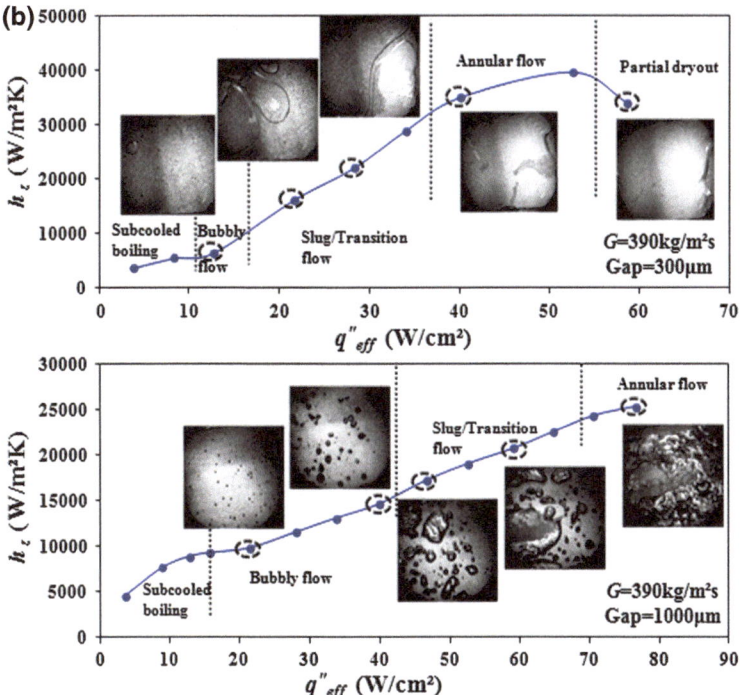

Fig. 5.9 continued

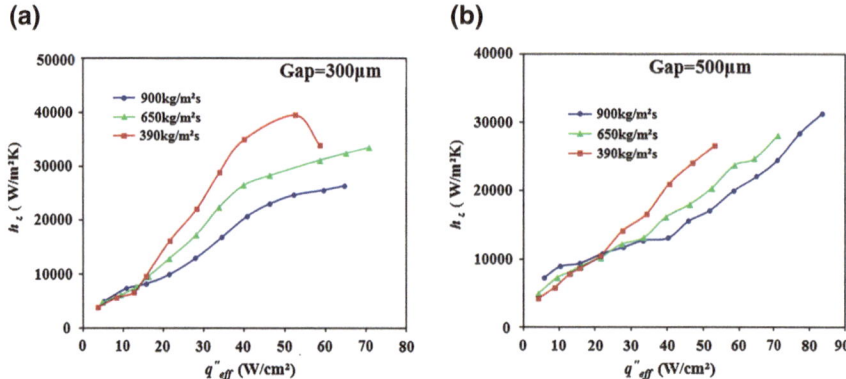

Fig. 5.10 Effect of mass flux on local heat transfer coefficients at microgap, **a** 300 μm, **b** 500 μm

Figure 5.10a and b show the effect of mass fluxes on local heat transfer coefficients at microgap, 300 and 500 μm. In single phase region, local heat transfer coefficient increases with the increase of mass flux. In contrast, local heat transfer coefficient decreases with the increase of mass flux in two-phase region. This result can be explained as- the vapor quality at the outlet is higher for a lower mass flux at a given heat flux. Because of these decreases of vapor quality, the local heat transfer coefficient decreases significantly at higher mass flux.

5.6 Uniformity of Wall Temperature

Microgap size effect on spanwise wall temperature uniformity over the heated surface at $G = 390$ kg/m²s and $q''_{eff} = 28$ W/cm² is shown in Fig. 5.11. The spanwise wall temperatures presented here are those measured near the exit at diode position 5, 10, 15, 20, and 25. It is noted from the figure that microgaps ranging from 100 to 360 μm maintain very uniform and low wall temperature, whereas 500 μm maintains uniform but slightly higher wall temperature. This is attributed to stable annular flow and thin film evaporative boiling over the heated surface in smaller gap. In the thin film evaporative boiling dominant region, higher boiling heat transfer is possible due to the thinner micro-layer [2–4]. As microlayer thickness decreases with decreasing gap size, the vaporization rate is increased and smaller microgaps maintain very uniform and low wall temperature. Microgaps range from 700 μm and above show a non-uniform wall temperature due to nucleate boiling dominant regime in these gaps. The smallest gap, 80 μm shows a sudden peak of wall temperature due to partial dryout by continuous thinning of thin liquid film.

Fig. 5.11 Microgap size effect on wall temperature (spanwise) uniformity curves

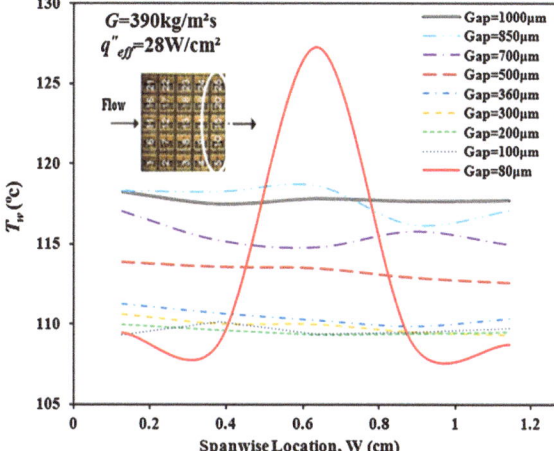

Fig. 5.12 Microgap size effect on wall temperature (streamwise) uniformity curves

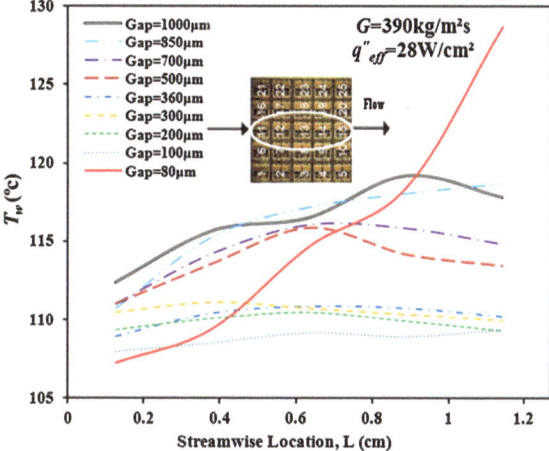

Microgap size effect on streamwise wall temperature uniformity over the heated surface at $G = 390$ kg/m^{2}s and $q''_{\mathrm{eff}} = 28$ W/cm^2 is shown in Fig. 5.12. The streamwise wall temperatures presented here are those measured at the middle row of diode position 11, 12, 13, 14, and 15. It is noted from figure that uniform and low wall temperature along streamwise location is maintained for microgaps range from 100 to 360 μm due to early establishment of confined slug/annular flow whereas 80 μm shows a increasing trend of wall temperature due to continuous thinning of thin film and early partial dryout. Microgaps range from 500 μm and above show a non-uniform wall temperature due to non-uniform distribution of boiling process over the heating surface.

References

1. Harirchian T, Garimella SV (2010) A comprehensive flow regime map for microchannel flow boiling with quantitative transition criteria. Int J Heat Mass Transf 53:2694–2702
2. Utaka Y, Okuda S, Tasaki Y (2009) Configuration of the micro-layer and characteristics of heat transfer in a narrow gap mini/micro-channel boiling system. Int J Heat Mass Transf 52:2205–2214
3. Han Y, Shikazono N (2010) The effect of bubble acceleration on the liquid film thickness in micro tubes. Int J Heat Fluid Flow 31:630–639
4. Han Y, Shikazono N, Kasagi N (2012) The effect of liquid film evaporation on flow boiling heat transfer in a micro tube. Int J Heat Mass Transf 55:547–555

Chapter 6
Surface Roughness Effect on Microgap Channel

Keywords Microgap channel · Surface roughness · Flow boiling · Flow visualization · Wall temperature uniformity · Heat transfer · Pressure drop

Understanding the influence of surface characteristics on flow boiling heat transfer behavior in microgap is necessary to enhance the performance of microgap heat sink. This chapter presents the heat transfer and pressure data of three different dimension microgap heat sinks with varying surface roughness collected during the experimental program. Experimental results are presented and discussed to investigate the influences of surface roughness on flow boiling heat transfer and pressure drop in microgap heat sink. High speed visualizations are shown to validate the explanation.

6.1 Flow Visualization

Flow visualizations of boiling processes taken from 500 μm gap at mass flux, $G = 390$ kg/m^{2}s and different imposed effective heat fluxes for three different microgap surface roughnesses, 0.6, 1.0, and 1.6 μm are illustrated in Fig. 6.1. It is observed for microgap of surface roughness, $R_a = 0.6$ μm from the Fig. 6.1a that only a few discrete bubbles nucleate, detach from and slide along the heating surface at the imposed heat flux 21 W/cm^2. No coalescence or confinement of bubbles is observed at this stage. With the increase of heat flux to 28 W/cm^2, the nucleating bubbles grow, confined in the gap, expand, and coalesce to form vapor slug. More coalescence and confinement of bubbles are observed as the imposed heat flux is increased to 40 W/cm^2. The expanded bubble occupies the entire microgap and slug flow regime transforms to confined annular flow regime as the imposed effective heat flux is raised to 59 W/cm^2. It is observed from the photo taken from the microgap of surface roughness, $R_a = 1.0$ μm from the Fig. 6.1b that lots of discrete bubbles nucleate, detach from and slide along the heating surface at the imposed heat flux 21 W/cm^2. It implies that the active bubble

T. Alam et al., *Flow Boiling in Microgap Channels*,
SpringerBriefs in Thermal Engineering and Applied Science,
DOI: 10.1007/978-1-4614-7190-5_6, © The Author(s) 2014

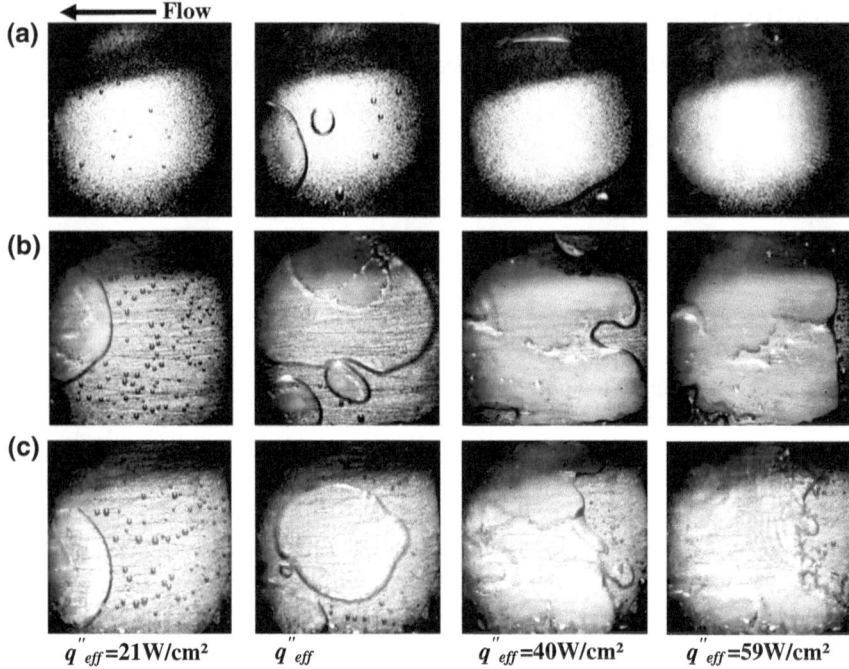

Fig. 6.1 Flow visualization of boiling process at various heat fluxes for 500 μm gap, $G = 390$ kg/m²s, **a** $R_a = 0.6$ μm, **b** $R_a = 1.0$ μm **c** $R_a = 1.6$ μm

nucleation site density increases with the increase of surface roughness. Coalescence and confinement of bubbles are observed near the exit of the heat sink at this stage. Bubbles coalesce more frequently in the microgap of surface roughness, $R_a = 1.0$ μm due to the higher nucleating bubble density as heat flux is increased to 28 W/cm². With the increase of heat flux to 40 and 59 W/cm², an annular flow regime with continuous coalescence of active nucleating bubbles and vigorous boiling is observed over the $R_a = 1.0$ μm microgap surface. Similar phenomena are observed with the further increase of the microgap of surface roughness, $R_a = 1.6$ μm as can be seen from the Fig. 6.1c.

Flow visualizations of boiling processes taken from 500 μm gap at increased mass flux, $G = 650$ kg/m²s and different imposed effective heat fluxes for three different microgap surface roughnesses, 0.6, 1.0, and 1.6 μm are illustrated in Fig. 6.2. The results in Figs. 6.1 and 6.2 indicate that the bubbles are smaller in size and depart at higher frequency from the heating surface at higher mass flux. Less activated nucleation sites on the heated surface are observed at higher mass flux as the fluid moves at a higher speed which interns causes the shorter time that fluid can be heated and requires more energy to activate the nucleation sites. Coalescence of bubbles followed by slug/annular flow regime is observed at higher imposed heat flux for this higher mass flux. In addition, effects of microgap surface

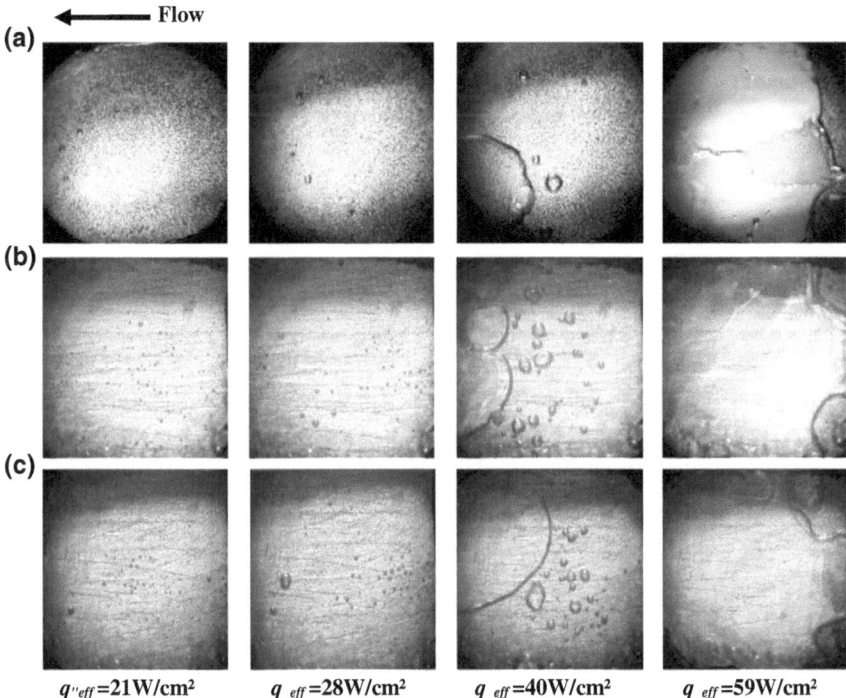

Fig. 6.2 Flow visualization of boiling process at various heat fluxes for 500 μm gap, $G = 650$ kg/m^{2}s, **a** $R_a = 0.6$ μm, **b** $R_a = 1.0$ μm, **c** $R_a = 1.6$ μm

roughness on bubble nucleation are also observed at higher mass flux and active bubble nucleation site density increases with the increase of surface roughness.

Flow visualizations of boiling processes taken from 300 μm gap at mass flux, $G = 390$ kg/m^{2}s and different imposed effective heat fluxes for three different microgap surface roughnesses, 0.6, 1.0, and 1.6 μm are illustrated in Fig. 6.3. It is noted by comparing the photos in Figs. 6.1 and 6.3 that at smaller microgap heat sinks, the effect of surface roughness is less marked. Confinement and coalescence of bubbles start at even low imposed heat flux 13 W/cm^2 and confined slug/annular flow dominant the boiling behavior at this 300 μm microgap heat sink.

6.2 Flow Boiling Curves

Surface roughness effects on boiling curves for different microgap sizes at mass flux, $G = 390$ kg/m^{2}s are presented in Fig. 6.4. Results are obtained for three different surface roughnesses, 0.6, 1.0, and 1.6 μm microgap heat sinks. The local wall temperatures presented here are those measured near the exit at diode position 15. It is noted from the Fig. 6.4a that the wall temperature exhibits a sudden change

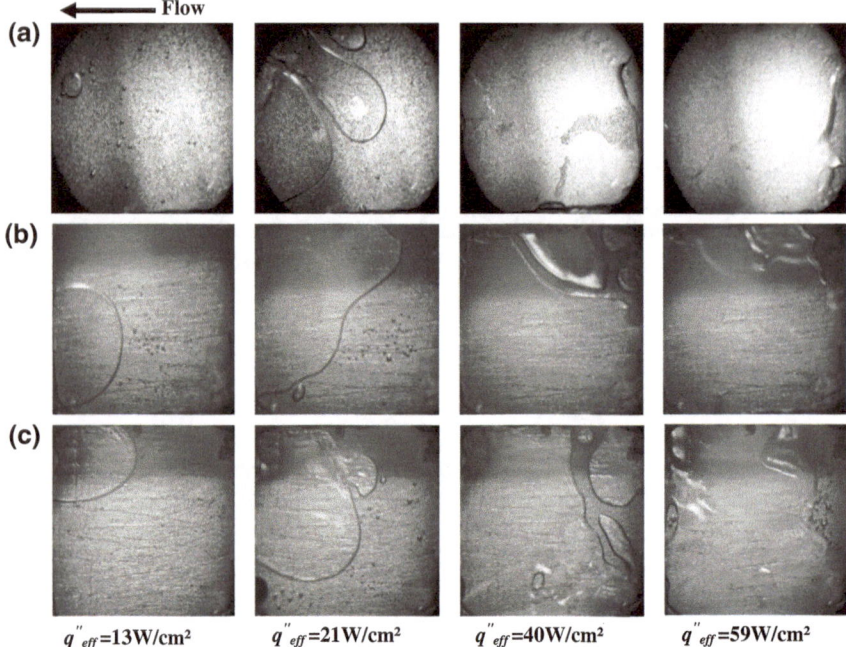

Fig. 6.3 Flow visualization of boiling process at various heat fluxes for 300 μm gap, $G = 390$ kg/m²s, **a** $R_a = 0.6$ μm, **b** $R_a = 1.0$ μm, **c** $R_a = 1.6$ μm

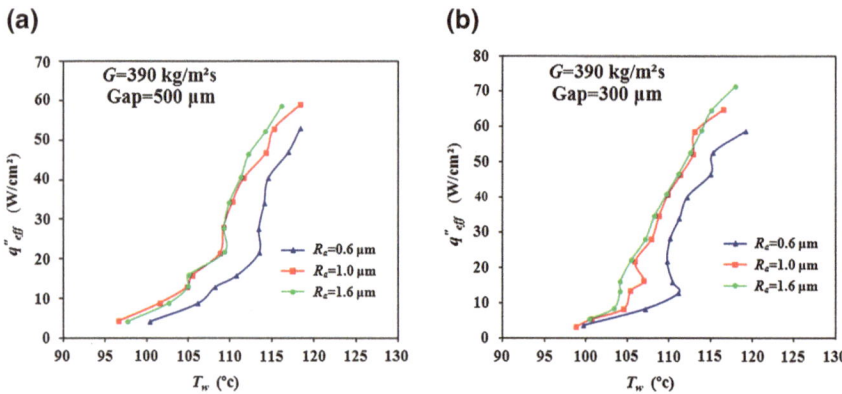

Fig. 6.4 Microgap surface roughness effect on boiling curves at mass flux, $G = 390$ kg/m²s for **a** gap = 500 μm, **b** gap = 300 μm

in slope from its single phase at the onset of nucleate boiling (ONB) for all three different roughness heat sinks. However, lower wall temperature is needed to commence boiling over the heated surface for rougher surface. It can also be seen from the figure that rougher surface microgap maintains lower chip wall

Fig. 6.5 Microgap surface roughness effect on boiling curves for 500 μm gap at mass flux, $G = 650$ kg/m^{2}s

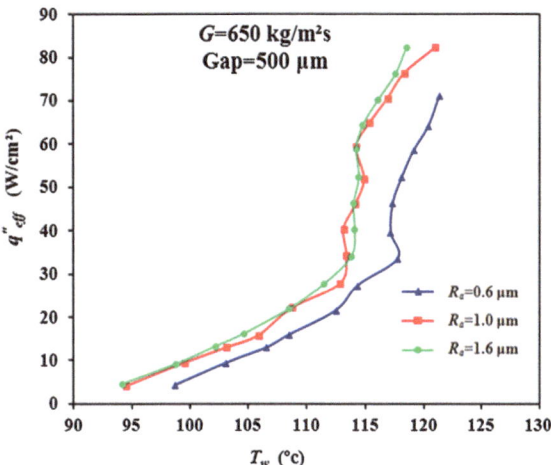

temperature at a fixed heat flux for both single and two-phase region for microgap of surface roughness 1 μm and above. Flow visualization in Fig. 6.1 revealed that this improvement of performance in rougher surface mainly due to presence of more active nucleation sites and high bubble population density. The surface roughness does not appear to have a substantial effect on boiling curve for microgap of surface roughness above 1 μm. All microgaps showed a similar trend. Boiling commences at wall superheats ranging from 8 to 14 K for microgap 500 μm; from 5 to 11 K for microgap 300 μm and from 4 to 9 K for microgap 200 μm. At larger dimension microgap heat sinks, the effect of surface roughness on the boiling curve is more marked. For 500 μm microgap heat sink, surface roughness has approximately 25–40 % higher effect than 300 μm and 200 μm microgap heat sink.

The influence of mass flux on boiling curve at microgap 500 μm can be compared between Figs. 6.4a and 6.5. Both mass fluxes showed a similar trend on surface roughness. However, lower wall temperature is needed to commence boiling over the heated surface for lower mass flux and boiling commences at wall superheats ranging from 8 to 14 K for mass flux 390 kg/m^{2}s and from 12 to 18 K for mass flux 650 kg/m^{2}s. These higher wall temperatures required for incipience of boiling at increased mass flux are due to the presence of larger temperature gradient near the wall, producing a thinner superheated liquid layer which suppress bubble nucleation as can be seen from flow visualizations in Figs. 6.1 and 6.2.

6.3 Local Flow Boiling Heat Transfer Coefficient Curves

Influence of surface roughness on local heat transfer coefficient curves for different microgap sizes at mass flux, $G = 390$ kg/m^{2}s are illustrated in Fig. 6.6. Results are obtained for three different microgap heat sinks of surface roughnesses, 0.6,

(a)   **(b)**

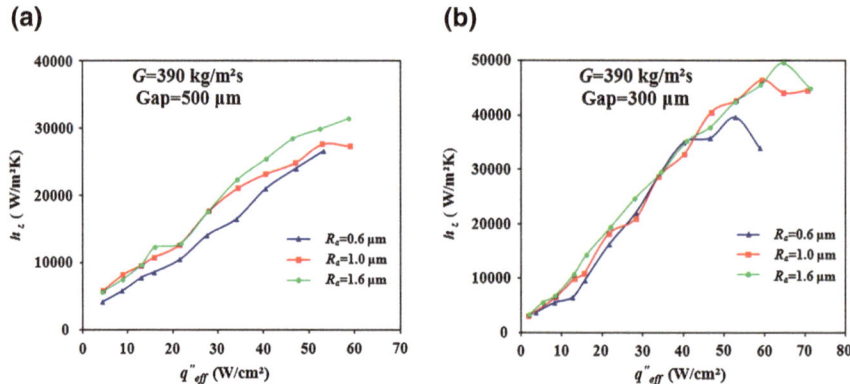

Fig. 6.6 Microgap surface roughness effect on local heat transfer coefficient curves at mass flux, $G = 390$ kg/m^{2}s for **a** gap $= 500$ μm, **b** gap $= 300$ μm

1.0, and 1.6 μm. The local heat transfer coefficients computed here are based on wall temperatures measured near the exit at diode position 15. It is noted from the Fig. 6.6a that surface roughness has effect on local heat transfer coefficient curve for both the single and two-phase regions and higher the surface roughness, higher the local heat transfer coefficients are achieved. This heat transfer coefficient enhancement at microgap of rougher surface may be due to increased wetted surface area in single phase region and increased bubble population density in two-phase region. At high heat flux, influence of surface roughness becomes significant on local heat transfer coefficient as the number of bubble nucleation sites, vapor bubble frequency and coalescence and confinement of bubbles increase with the increase of heat flux. Similar trend is observed for other microgap sizes.

The influence of mass flux on local heat transfer coefficient curve at microgap 500 μm can be compared between Figs. 6.6a and 6.7. Both mass fluxes showed a similar trend on surface roughness.

6.4 Pressure Drop Curves

The pressure drop curves for three different surface roughnesses, 0.6, 1.0, and 1.6 μm at mass flux, $G = 390$ kg/m^{2}s and microgap size 300 μm are illustrated in Fig. 6.8. Pressure drop is measured between the two upstream and downstream manifolds and sudden contraction and expansion losses are corrected. Pressure drop is independent of heat flux at low heat flux until the boiling starts and pressure drop increases gradually with the increase of heat flux due to the dominance of acceleration effect of vapor content as boiling starts. No significant influence of surface roughness on pressure drop curve is observed for both the single and two-phase regions below approximately 50 W/cm^2. For heat fluxes

Fig. 6.7 Microgap surface roughness effect on local heat transfer coefficient curves for 500 μm gap at mass flux, $G = 650$ kg/m²s

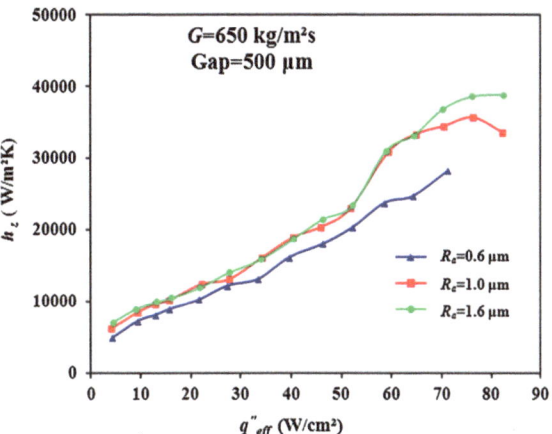

Fig. 6.8 Microgap surface roughness effect on pressure drop curves for 300 μm gap at mass flux, $G = 390$ kg/m²s

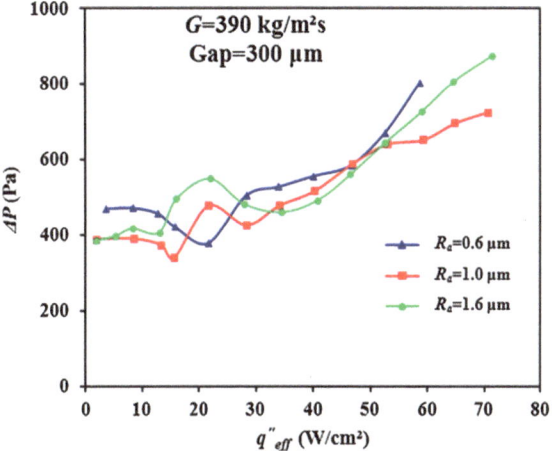

above 50 W/cm², there appears to be a significant effect of surface roughness on pressure drop curve. Similar trend is observed for higher mass flux and higher gap size as shown in Fig. 6.9.

6.5 Uniformity of Wall Temperature

Figure 6.10 presents the trend of local wall temperature versus the spanwise location taken from 500 μm gap at mass flux, $G = 390$ kg/m²s, for three different microgap surface roughnesses, $R_a = 0.6$, 1.0, and 1.6 μm. The spanwise local wall temperatures presented here are those measured near the inlet, middle, and exit of the test section as marked in Fig. 6.10. Rougher surface maintains lower wall

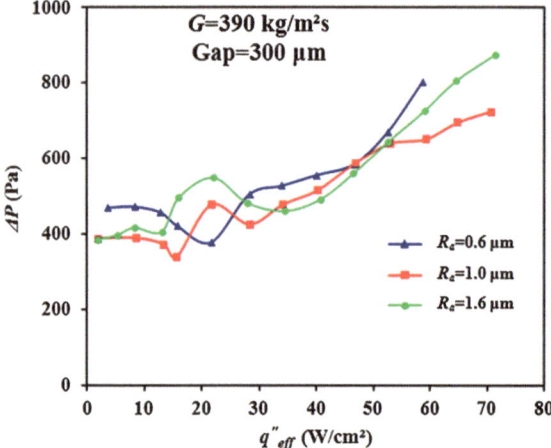

Fig. 6.9 Microgap surface roughness effect on pressure drop curves for 500 μm gap at mass flux, $G = 650$ kg/m^{2}s

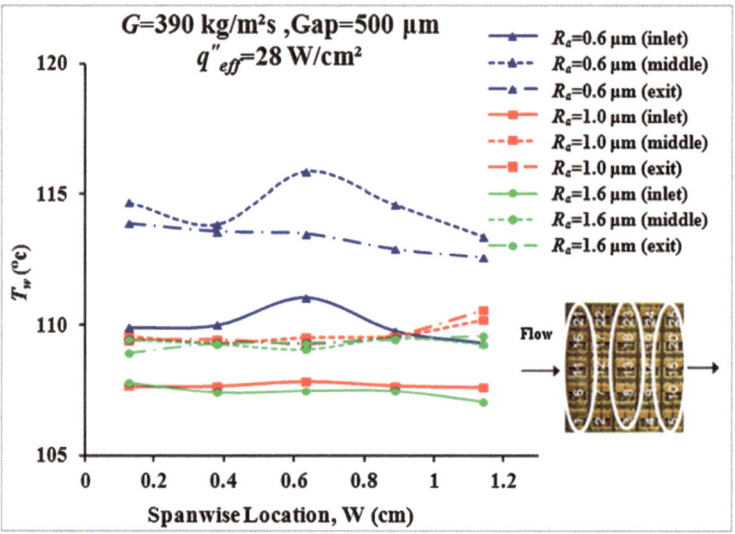

Fig. 6.10 Microgap surface roughness effect on wall temperature uniformity curves for 500 μm gap, $G = 390$ kg/m^{2}s at heat fluxes, $q''_{\text{eff}} = 28$ W/cm^2

temperature over the heated surface. In addition, a large variation of wall temperature along both spanwise and streamwise location is observed for microgap of surface roughness, $R_a = 0.6$ μm and these variations of wall temperature decrease with the increase of microgap surface roughness. A variation of spanwise wall temperature is observed ranging from 1.3 to 2.5 K for $R_a = 0.6$ μm and from 0.5

to 1 K for $R_a = 1.0$ and 1.6 µm. In addition, a variation of streamwise wall temperature is observed ranging from 3.5 to 4.8 K for $R_a = 0.6$ µm and from 1.7 to 2.9 K for $R_a = 1.0$ and 1.6 µm. This large variation of wall temperature along both spanwise and streamwise location for $R_a = 0.6$ µm is due to non-uniform distribution of boiling process over the heated surface [1]. Flow visualization from Fig. 6.1 revealed that the active bubble nucleation site density increases with the increase of surface roughness. Bubbles coalesce more frequently to form slug/annular regime in the microgap of higher surface roughness due to the higher nucleating bubble density which result uniform and low wall temperature. A similar trend was observed for all heat fluxes, mass fluxes, and microgaps sizes.

Reference

1. Alam T, Lee PS, Yap CR, Jin LW, Balasubramanian K (2012) Experimental investigation and flow visualization to determine the optimum dimension range of microgap heat sinks. Int J Heat Mass Transf 55:7623–7634

Chapter 7
Two-Phase Microgap Channel in Mitigating Flow Instabilities and Flow Reversal

Keywords Microgap channel · Microchannel · Surface roughness · Flow boiling · Flow visualization · Flow instabilities · Flow reversal

Flow boiling instabilities induce mechanical vibration in the system and deteriorate the heat transfer performances, for example—premature dryout, critical heat flux limitation, etc. The two-phase microgap heat sink has novel potential to mitigate these undesirable flow boiling instabilities and flow reversal issues inherent with two-phase microchannel heat sink. This chapter presents the comparison of instabilities in microgap heat sink with conventional straight microchannel heat sink at the beginning of the chapter. The comparison is done with the same footprint, the same inlet mass flux, and the same effective heat flux supplied based on the footprint. In the later section, this chapter presents the influencing parameters that affect the instabilities in microgap heat sink. High speed visualizations are shown to validate the experiment results and explanation.

7.1 Comparison of Instabilities Between Microgap and Microchannel

The present flow boiling experiments with high speed flow visualizations are performed for microgap and microchannel heat sink to compare the instability characteristics.

Sequential flow pattern in the microchannel at $G = 420$ kg/m^{2}s and $q''_{\text{eff}} = 18$ W/cm^2 is demonstrated in Fig. 7.1. The first frame at t shows the start of the generation of bubble. As time progress, the bubbles start to grow and occupy almost the entire microchannel width as shown at $(t + 0.0016$ s$)$. These bubbles expand further to form slug/annular flow pattern only in axial direction as vapor growth phase is limited in the radial direction as shown at $(t + 0.0034$ s$)$. In contrast, the nucleating bubble has room to expand both spanwise and downstream

T. Alam et al., *Flow Boiling in Microgap Channels*,
SpringerBriefs in Thermal Engineering and Applied Science,
DOI: 10.1007/978-1-4614-7190-5_7, © The Author(s) 2014

Fig. 7.1 Sequential images of flow regime development at microchannel, $G = 420$ kg/m²s, $q''_{\mathrm{eff}} = 18$ W/cm²

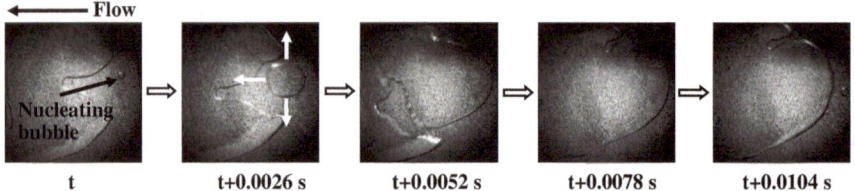

Fig. 7.2 Sequential images of flow regime development at microgap, $G = 420$ kg/m²s, $q''_{\mathrm{eff}} = 29$ W/cm²

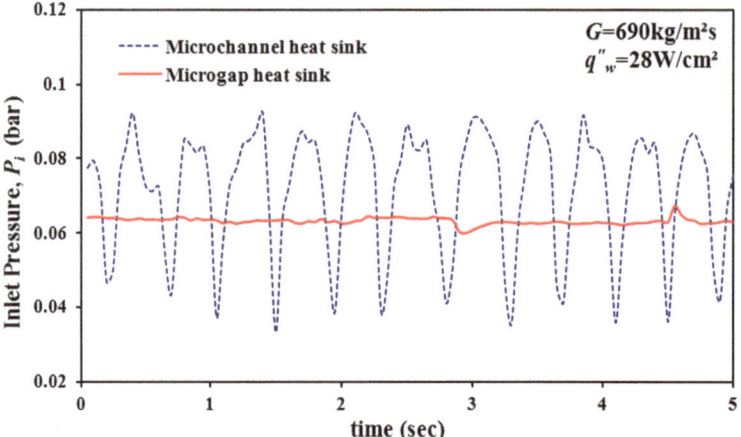

Fig. 7.3 Comparison of pressure instability in micochannel and microgap channel at $G = 690$ kg/m²s and $q''_{\mathrm{w}} = 28$ W/cm²

instead of being forced upstream in microgap heat sink as demonstrated the sequential flow pattern of microgap heat sink in Fig. 7.2.

Inlet pressure fluctuation for microchannel and microgap heat sink at a given wall heat flux has been demonstrated in Fig. 7.3. It is revealed that microchannel

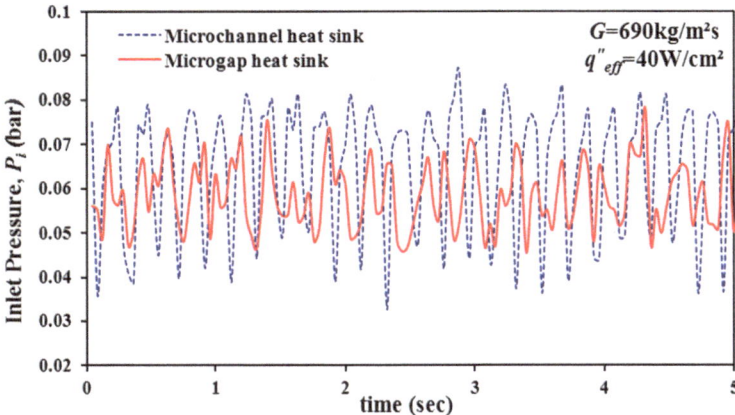

Fig. 7.4 Comparison of pressure instability in micochannel and microgap at $G = 690$ kg/m^{2}s and $q''_{\text{eff}} = 40$ W/cm^2

shows more inlet pressure fluctuation than microgap heat sink at same wall heat flux condition. In a small sized channel, the vapor growth phase is limited in the radial direction because of the hydraulic diameter as shown in Fig. 7.1. Only the axial direction allows vapor growth when boiling occurs. On the contrary, in microgap heat sink, the vapor generated has room to expand both spanwise and downstream instead of being forced upstream as presented in Fig. 7.2 which minimizes the pressure fluctuation.

Further, instead of wall heat flux if inlet pressure fluctuation is plotted at a given effective heat flux as is done in Fig. 7.4, it is also seen that for a fixed heat dissipation rate from the chip, pressure fluctuation is smaller for microgap compared to microchannel heat sink.

Figure 7.5 compares the wall temperature fluctuation at diode position 15 between microgap and microchannel at wall heat flux 28 W/cm^2 and mass flux 690 kg/m^{2}s. Microgap maintains much lower wall temperature fluctuation than microchannel as can be seen from the figure. According to Consolini and Thome [1], temperature fluctuation may arise because of either the local change in flow temperature associated with the pressure fluctuations due to the bubble growth, expansion, and flushing process or the cyclical variations in the heat transfer mechanisms. In Fig. 7.3, it can be seen that pressure fluctuation is much higher for microchannel than microgap. Microgap maintains a stable flow regime throughout the heat sink which incorporates lower pressure and consequently lowers wall temperature fluctuation. In contrast, microchannel shows an unstable boiling characteristic due to very high fluctuation of pressure which incorporates a high wall temperature fluctuation over the heat sink.

Further, if wall temperature fluctuation is plotted at a given effective heat flux instead of wall heat flux as is done in Fig. 7.6a and b; it is seen from Fig. 7.6a that for a fixed heat dissipation rate, 40 W/cm^2 from the chip, wall temperature fluctuation is comparatively smaller for microgap compared to microchannel heat

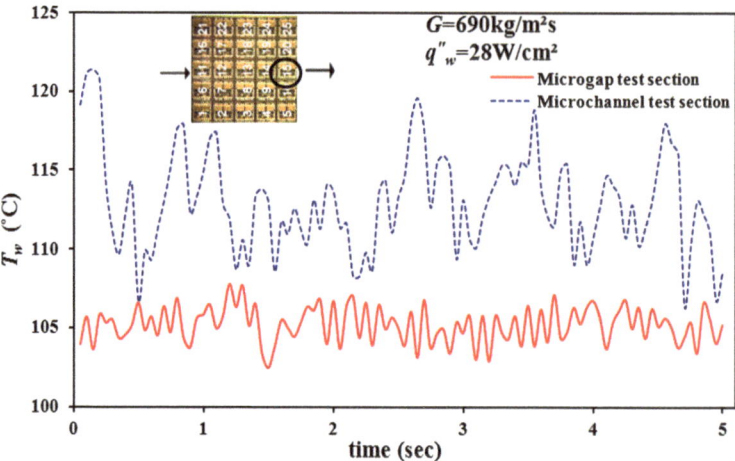

Fig. 7.5 Comparison of local wall temperature fluctuation between microchannel and microgap channel at diode position 15

sink. However, mean wall temperature is higher for microgap compare to microchannel as early establishment of slug/annular flow and consequent rise of vapor quality in microchannel of very small diameter attributes the better heat transfer performance at low heat flux. As the heat flux increases to 51 W/cm^2 in Fig. 7.6b, the wall temperature fluctuation still remain low for microgap and mean wall temperature also goes down for this heat sink configuration due to confined slug and annular boiling dominance and consequent thin film evaporation in microgap heat sink at higher heat flux region.

7.2 Influencing Factors of Instabilities in Microgap Heat Sink

The flow boiling instability characteristics in microgap heat sink over a range of gap sizes, mass fluxes, heat fluxes, and surface roughnesses with high speed flow visualizations are performed for better fundamental understanding and to identify the influencing parameter of instabilities in microgap heat sink.

7.2.1 Microgap Size Effects

Figure 7.7 demonstrates the periodic flow pattern of 1,000 μm gap at $G = 390$ kg/m^{2}s and $q''_{\text{eff}} = 71$ W/cm^2. It can be seen from the figure that time t corresponds to the initiation of boiling, when multiple small bubbles start to develop. As time

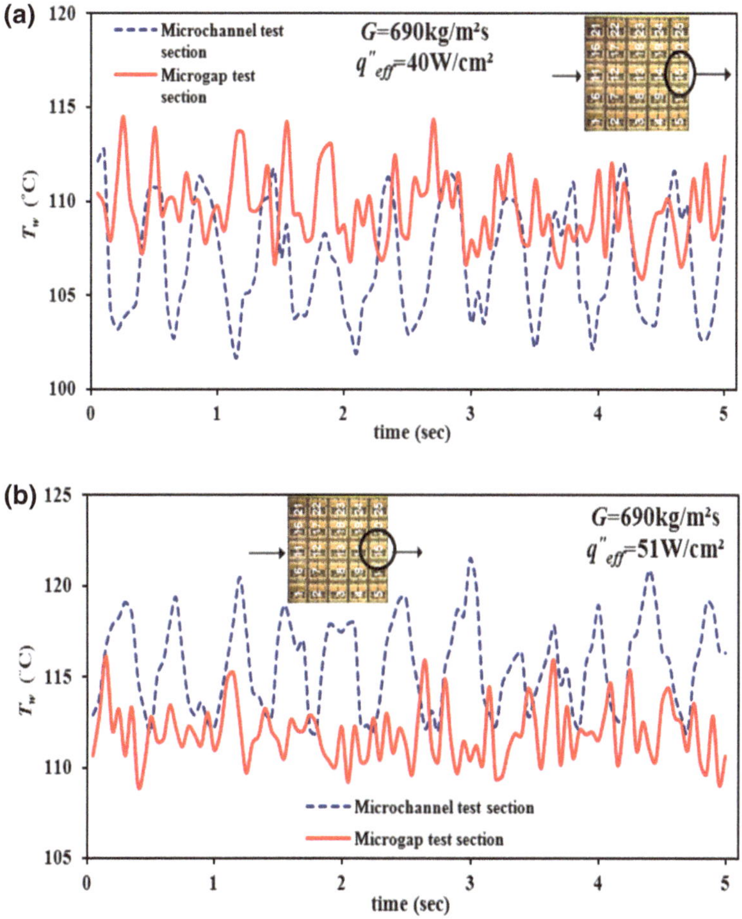

(a)

G=690kg/m²s
q''_{eff}=40W/cm²

- - - Microchannel test section
— Microgap test section

T_w (°C)

(b)

G=690kg/m²s
q''_{eff}=51W/cm²

- - - Microchannel test section
— Microgap test section

T_w (°C)

time (sec)

Fig. 7.6 Comparison of local wall temperature fluctuation between microchannel and microgap at diode position 15 at heat flux, **a** $q''_{eff} = 40$ W/cm², **b** $q''_{eff} = 51$ W/cm²

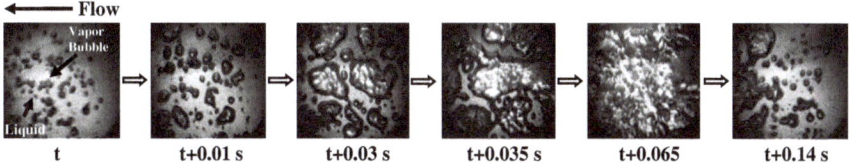

Flow

Vapor Bubble

Liquid

t   t+0.01 s   t+0.03 s   t+0.035 s   t+0.065   t+0.14 s

Fig. 7.7 Sequential images of flow regime development at 1,000 μm gap, $G = 390$ kg/m²s, $q''_{eff} = 71$ W/cm²

progress, the bubbles grow due to evaporation, as liquid is converted to vapor at the microgap wall at ($t + 0.01$ s). These multiple bubbles grow further with time and coalesce with each other to build vapor slugs where some thin layer of liquid

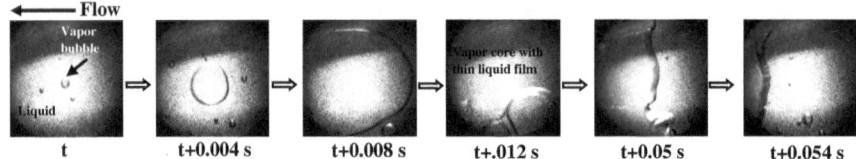

Fig. 7.8 Sequential images of flow regime development at 500 μm gap, $G = 390$ kg/m²s, $q''_{eff} = 40$ W/cm²

are seen trapped between the slugs at $(t + 0.03$ s). With the further progression of time at $(t + 0.035$ s), as the slugs expand, the liquid between the slugs begin to shrink as the churn\annular flow regime occupies the microgap channel shown at time $(t + 0.065$ s). At time $(t + 0.14$ s), this flow regime move downstream and new bubble start to generate over the surface to repeat the process.

In contrast to the 1,000 μm gap, a 500 μm gap at $G = 390$ kg/m²s and $q''_{eff} = 40$ W/cm² exhibits a different flow boiling behavior as shown in periodic flow pattern in Fig. 7.8. As shown at time t, only a few bubbles start to generate over the heated surface and one of these bubbles reach the size of the microgap depth. This bubble then starts to expand in all direction and to push back the other generated bubble over the heated surface as can be seen at $(t + 0.004$ s). The bubble expands further until confinement in the microgap channel at $(t + 0.008$ s). As time progress, other growing bubbles from upstream start to grow and merge with the confined bubble at $(t + 0.012$ s). So, the microgap heated wall is constantly covered with confined bubbles and thin film evaporation occurs throughout the liquid vapor interface. Thereafter, the confined bubble finally moves downstream from the heated wall under the effect of liquid flow at $(t + 0.05$ s) and new bubbles start to grow such that the process is repeated.

A sequential flow pattern of 80 μm gap at $G = 390$ kg/m²s and $q''_{eff} = 16$ W/cm² is demonstrated in Fig. 7.9. The first frame at t shows the start of the generation of bubble and unlike larger gap, a single bubble is sufficient to create confinement. As time progress, the bubble starts to expand in explosive manner as shown at $(t + 0.006$ s) and bubble confinement is established within the microgap only at $(t + 0.012$ s). The confined bubble then moves downstream and new bubble generates over the microgap heated wall at $(t + 0.035$ s) and the process is repeated.

Fig. 7.9 Sequential images of flow regime development at 80 μm gap, $G = 390$ kg/m²s, $q''_{eff} = 16$ W/cm²

From the above investigations, it is clear that the procedures of flow regime development were different for all three gap dimensions. No confinement effect on bubble growth was observed for 1,000 μm gap and bubble coalescence play a very important role in governing the entire flow regime. On the other hand, both 500 and 80 μm and show confinement effect on bubble growth. However, a single nucleation is sufficient to create confinement all over the heated surface for 80 μm whereas multiple bubbles are needed for 500 μm gap. Moreover, smallest gap exhibits confinement at lowest heat flux as well as shortest periodic sequence.

Figure 7.10 shows the microgap size effect on inlet pressure fluctuation at $G = 390$ kg/m^{2}s and $q''_{\text{eff}} = 28$ W/cm^2 as a function of time. It is observed from the figure that the amplitude of inlet pressure oscillation decreases with increasing gap size. In this study, standard deviation of the fluctuating inlet pressure data is used as a measure of the amplitude of fluctuation. Based on this, the pressure fluctuation at 80 μm gap is found to be approximately 35 and 50 % higher than 360 and 1,000 μm gap respectively. This high fluctuation is due to the occurrence of vapor flow reversal in smaller gap. This flow reversal indicates the unstable boiling phenomenon at smaller gap due to expanding bubble during confinement and also premature partial dryout. Simultaneous flow boiling visualization inside microgap channel shows that these fluctuations are caused by flow alternation between liquid, two-phase and vapor flow. During flow boiling in small microgap, bubble grows from nucleation site, expand both upstream and downstream and occupy the entire gap. This downstream expansion push the liquid front back leading to reverse flow. Thus, liquid film is formed by reverse flow. After liquid film is formed on the wall, liquid film thickness quickly decreases with large

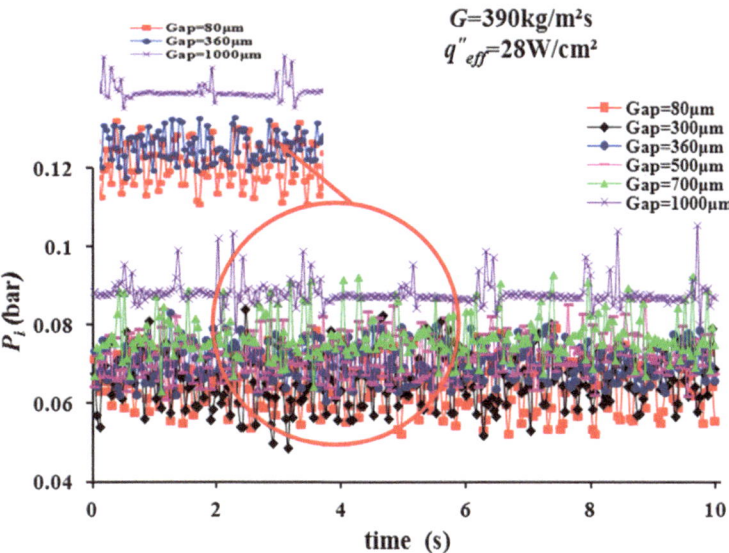

Fig. 7.10 Microgap size effect on inlet pressure fluctuation

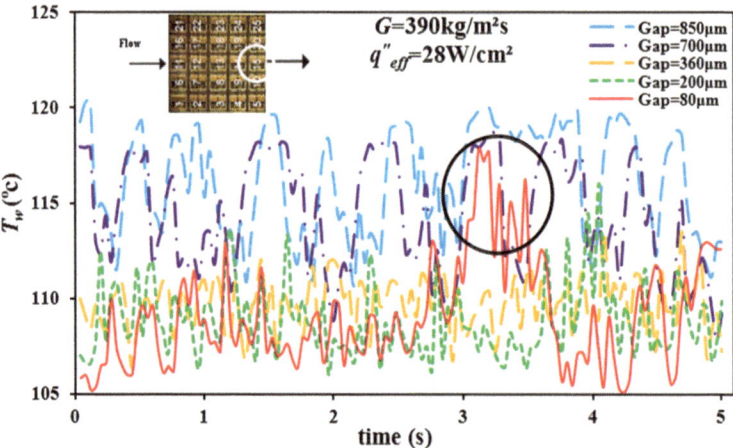

Fig. 7.11 Microgap size effect on wall temperature (diode position 15) fluctuation curves

fluctuation due to strong evaporation [2]. Hence, partial dryout period occurs due to thin liquid film evaporation in the confined annular flow. Then incoming subcooled liquid condenses vapor and a new cycle starts with bubble nucleation and repeat in cycles. Sequential flow pattern for different microgap sizes (Figs. 7.7, 7.8 and 7.9) shows that this cycle for smaller microgap lasts much shorter period and achieves at low heat flux than the case of larger microgap as thin film thickness decreases with decreasing gap size and therefore higher fluctuations in smaller gap are observed.

Figure 7.11 illustrates the microgap size effect on local wall temperature fluctuation curves at $G = 390$ kg/m²s and $q_{\text{eff}}'' = 28$ W/cm² as a function of time. From figure, it is seen that magnitude of wall temperature and wall temperature fluctuation are lower for smaller gap compare to larger gap except 80 μm. The local wall temperature fluctuation at 700 μm gap is approximately 45 % higher than 200 μm. High speed visualization of flow boiling in microgap heat sink revealed that nucleate boiling is the dominant heat transfer mechanism in larger gap whereas confined slug/annular flow and thin film evaporation are dominant for smaller gap. A cycle of bubble nucleation, detachment of bubble and surface rewetting over the heated surface may be the reason for higher fluctuation in larger gap. In contrast, a stable annular thin film regime and thin film evaporation may reason for lower magnitude of wall temperature and wall temperature fluctuation in smaller gap. At 80 μm gap, a sudden peak of wall temperature may be due to the unstable annular regime with periodic dryout and rewetting at this heat flux.

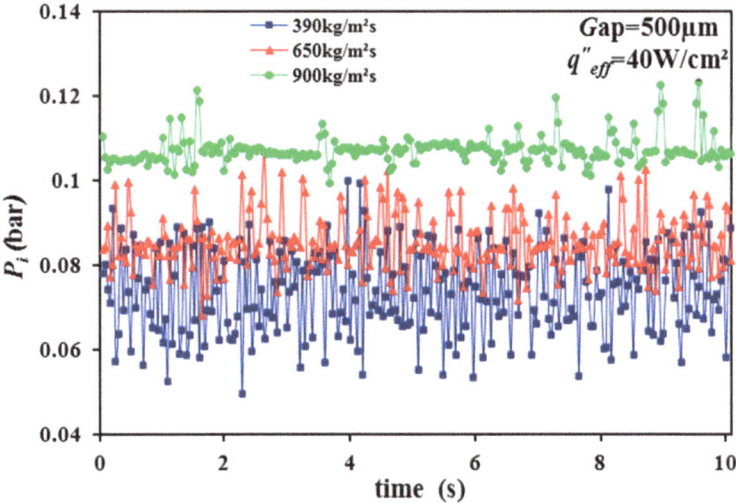

Fig. 7.12 Inlet pressure fluctuation for 500 μm gap at various mass fluxes

7.2.2 Mass Flux Effects

Figure 7.12 illustrates the instabilities in the inlet pressure at various mass fluxes taken from 500 μm gap and 40 W/cm^2 as a function of time. It indicates that the instability in the inlet pressure is lower in amplitude for the higher mass flux for a fixed heat flux. The pressure fluctuation increases with increasing vapor content. For the same heat flux, as the mass flux increases, the vapor content decreases as shown in Fig. 7.13, leading to the trends shown in Fig. 7.12.

Wall temperature fluctuation at various mass fluxes taken from 500 μm gap at 40 W/cm^2 is illustrated at Fig. 7.14. The results in Fig. 7.14 indicate that magnitude of wall temperature increases with increasing mass flux. However, large wall temperature fluctuation is observed as mass flux decreases from 900 to 650 kg/m^{2}s. With the increase of mass flux, the single phase convective heat transfer increases, which suppress bubble nucleation, resulting lower wall temperature fluctuation. A decrement of wall temperature fluctuation is observed as mass flux further decreases from 650 to 390 kg/m^{2}s may be due to the establishment of confined annular flow over the heated surface.

7.2.3 Heat Flux Effects

Flow visualization of boiling processes at different imposed heat fluxes taken from 360 μm gap at $G = 390$ kg/m^{2}s are shown in Fig. 7.15. It is noted from Fig. 7.15a that numerous discrete bubbles nucleate, detach from and slide along the heating

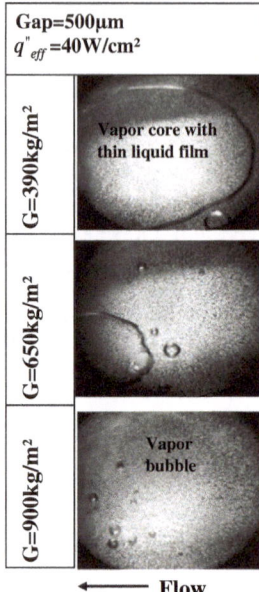

Fig. 7.13 Flow visualization of boiling process for 500 μm gap, $q''_{\text{eff}} = 40$ W/cm² at various mass fluxes

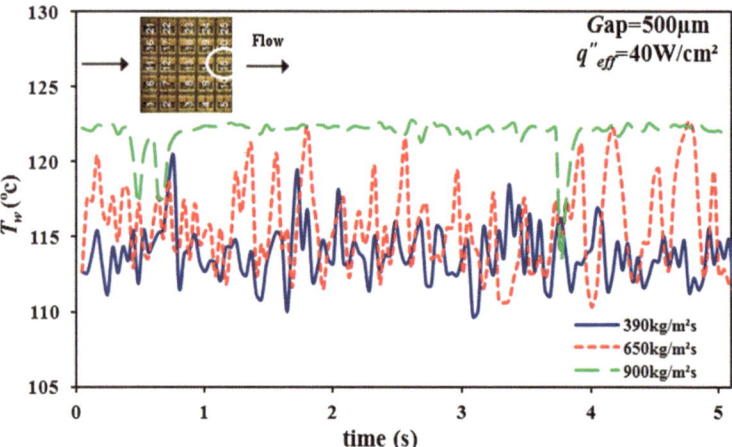

Fig. 7.14 Wall temperature (diode position 15) fluctuation for 500 μm gap at various mass fluxes

surface at the imposed heat flux 21 W/cm². It implies that many bubble nucleation sites are activated. During sliding along the heating surface, detached bubbles gain heat from the heated surface and expand slightly as can be seen at the exit of the gap. As the imposed heat flux is increased to 28 W/cm² as shown in Fig. 7.15b, the

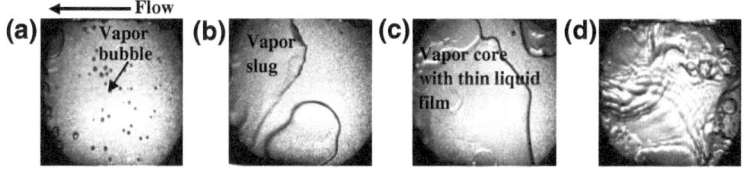

Fig. 7.15 Flow visualization of boiling process for 360 μm gap, $G = 390$ kg/m^{2}s at heat fluxes **a** $q''_{\text{eff}} = 21$ W/cm^2, **b** $q''_{\text{eff}} = 28$ W/cm^2, **c** $q''_{\text{eff}} = 40$ W/cm^2, **d** $q''_{\text{eff}} = 59$ W/cm^2

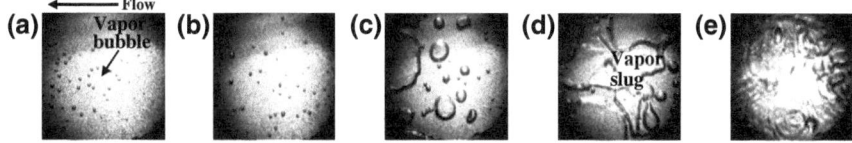

Fig. 7.16 Flow visualization of boiling process for 700 μm gap, $G = 390$ kg/m^{2}s at heat fluxes **a** $q''_{\text{eff}} = 21$ W/cm^2, **b** $q''_{\text{eff}} = 28$ W/cm^2, **c** $q''_{\text{eff}} = 40$ W/cm^2, **d** $q''_{\text{eff}} = 59$ W/cm^2, **e** $q''_{\text{eff}} = 70$ W/cm^2

nucleating bubbles grow and confined in the gap. These confined bubbles then expand and coalesce to form vapor slug. With the further increase of heat flux to 40 W/cm^2, the liquid in the slugs between the bubbles begins to shrink as the expanded bubble occupies the entire microgap to develop a confined annular flow pattern as presented in Fig. 7.15c. A vigorous boiling followed by partial dryout in the microgap is observed as the imposed heat flux is raised to 59 W/cm^2 as shown in Fig. 7.15d.

Figure 7.16 illustrates the flow visualization of boiling processes at different imposed heat fluxes taken from 700 μm gap at $G = 390$ kg/m^{2}s. At the imposed heat flux 21 W/cm^2, discrete bubbles nucleate over the heated surface, detach, and move downstream. Like 360 μm, detached bubbles gain heat from surface and expand slightly during moving downstream but bubble departure diameter is much smaller than 360 μm for 700 μm at the exit of microgap. Bubbles grow slightly bigger before departing from the heated surface as the imposed heat flux is increased to 28 W/cm^2. Coalescence of bubbles is rarely observed at this stage. As heat flux is raised further to 40 W/cm^2, growth rates of bubbles are much increased and the bubbles in the gap coalesce to form bigger bubble before dispatch from the heated surface. Bubbly flow regime transforms into slug flow regime at even higher imposed heat flux to 59 W/cm^2 as a result of high coalesce of bubbles before travelling downstream. With the further increase of heat flux to 70 W/cm^2, a vigorous boiling followed by partial dryout is observed for this gap.

Figure 7.17 shows the inlet pressure fluctuation curves taken from 360 μm gap at 390 kg/m^{2}s at various heat fluxes. It is apparent from the figure that as the heat flux increases, the inlet pressure fluctuations increases. With the aid of flow visualization, it is seen that this large inlet pressure fluctuation at higher heat flux could be attributed due to the shape of vapor slug. A cycle of vapor slug/chunk

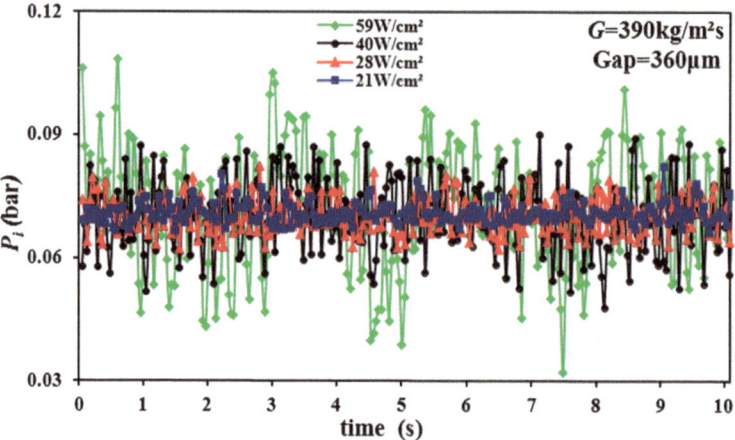

Fig. 7.17 Inlet pressure fluctuation curves for 360 μm gap at various heat fluxes

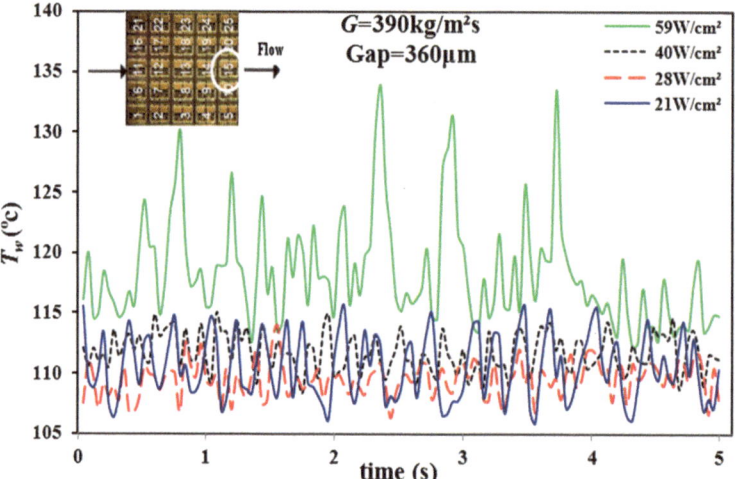

Fig. 7.18 Wall temperature (diode position 15) fluctuation curves for 360 μm gap at various heat fluxes

growing and shrinking within the microgap caused unstable flow condition and flow reversal which intern show high pressure fluctuation.

Wall temperature fluctuation at various imposed heat flux taken from 360 μm gap at 390 kg/m²s is illustrated at Fig. 7.18. A moderate wall temperature fluctuation is observed at heat flux, 21 W/cm² due to localized nucleate boiling. A decrease in wall temperature fluctuation is observed by approximately 37 and 28 %, as the heat fluxes increase to 28 and 40 W/cm² respectively due to thin film evaporation during confined vapor slug or annular flow. With the further increase of heat flux to 59 W/cm², a large fluctuation and sudden peak of wall temperature are observed due to vigorous boiling and partial dryout in microgap.

7.2.4 Surface Roughness Effects

Figure 7.19 shows the surface roughness effect on inlet pressure fluctuation curves for different microgap sizes as a function of time. Results are obtained for three different surface roughnesses, $R_a = 0.6$, 1.0, and 1.6 μm microgap heat sinks. In

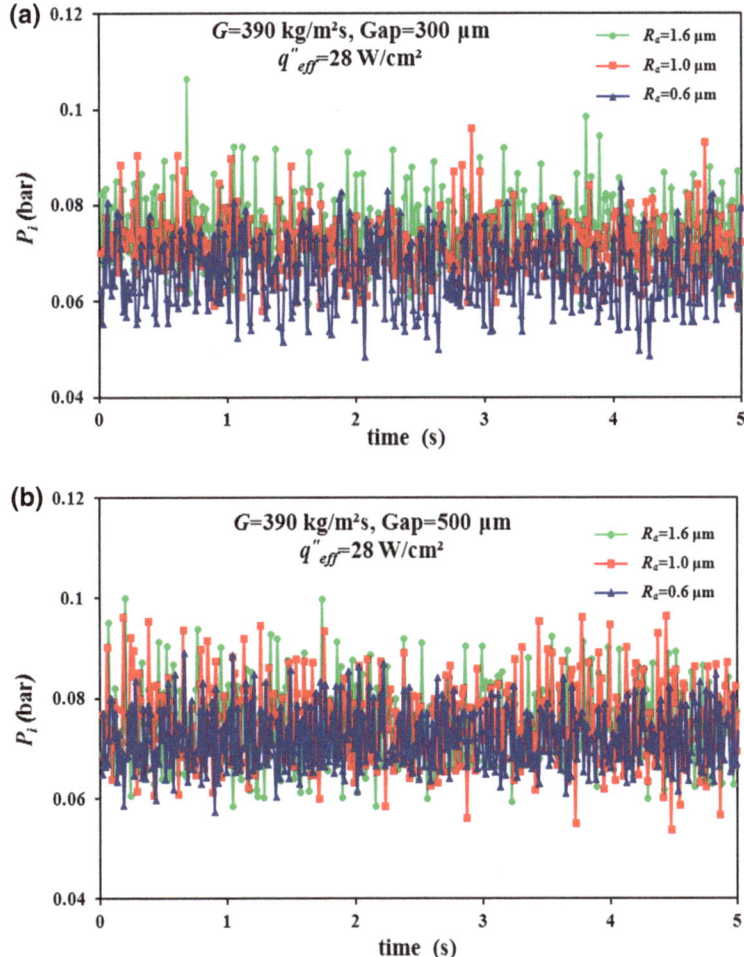

Fig. 7.19 Microgap surface roughness effect on inlet pressure fluctuation curves at $G = 390$ kg/m²s and $q''_{eff} = 28$ W/cm² for **a** gap = 300 μm, **b** gap = 500 μm

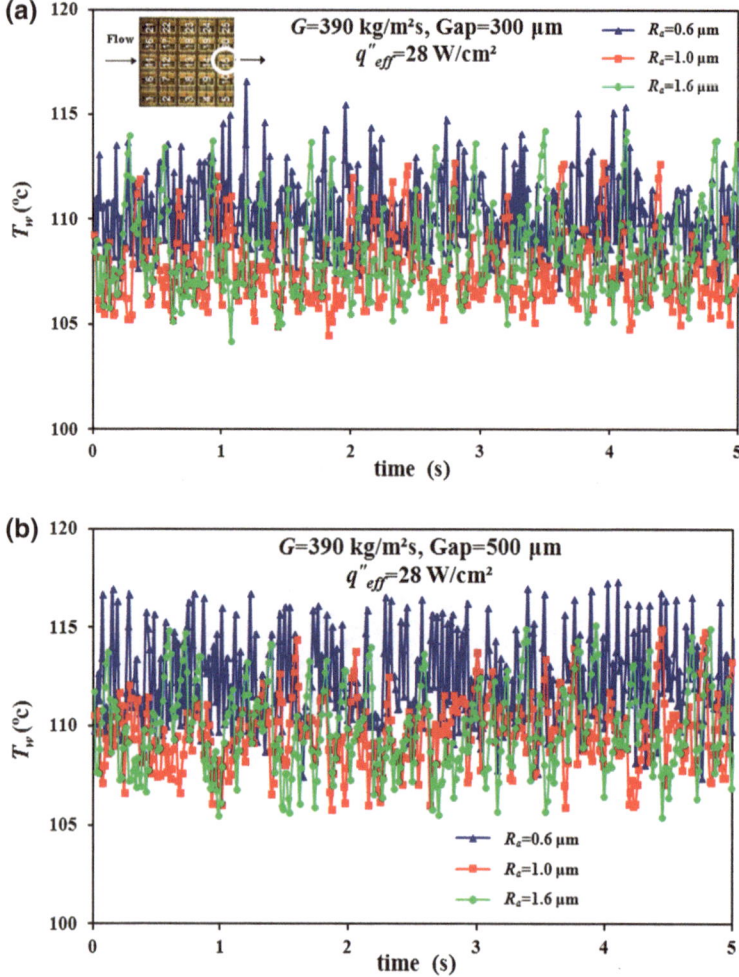

Fig. 7.20 Microgap surface roughness effect on wall temperature (diode position 15) fluctuation curves at $G = 390$ kg/m^{2}s and $q''_{\text{eff}} = 28$ W/cm^2 for **a** gap $= 300$ μm, **b** gap $= 500$ μm

this study, standard deviation of the fluctuating inlet pressure data is used as a measure of the amplitude of fluctuation. Based on this, the inlet pressure fluctuation for $R_a = 0.6$ μm is found to be approximately equal to $R_a = 1.0$ μm at 300 μm microgap as shown in Fig. 7.19a. Pressure fluctuation is attributed due to the shape of vapor slug. A cycle of vapor slug/chunk growing and shrinking within the microgap caused unstable flow condition and flow reversal which intern show pressure fluctuation [3]. Flow visualization in Fig. 6.3 showed that at smaller microgap heat sinks, the effect of surface roughness is less marked and similar boiling behavior is observed. However, the inlet pressure fluctuation for $R_a = 0.6$ μm is found to be approximately 20 % lower than $R_a = 1.0$ μm at

500 μm microgap as can be seen from Fig. 7.19b. Thus, it can be concluded from the above observations that the influence of surface roughness on inlet pressure fluctuation increases with increasing microgap sizes.

Figure 7.20 shows the surface roughness effect on local wall temperature fluctuation curves for different microgap sizes as a function of time. Results are obtained for three different microgap heat sinks of surface roughnesses, $R_a = 0.6$, 1.0, and 1.6 μm. The local wall temperatures presented here are those measured near the exit at diode position 15. From figure, it is seen that magnitude of wall temperature is slightly higher for microgap of surface roughness, $R_a = 0.6$ μm compare to microgap of surface roughness, $R_a = 1.0$ and 1.6 μm. However, wall temperature fluctuations are independent of surface roughness for microgaps.

References

1. Consolini L, Thome JR (2009) Micro-channel flow boiling heat transfer of R-134a, R-236fa, and R-245fa. Microfluid Nanofluid 6:731–746
2. Han Y, Shikazono N (2010) The effect of bubble acceleration on the liquid film thickness in micro tubes. Int J Heat Fluid Flow 31:630–639
3. Alam T, Lee PS, Yap CR, Jin LW, Balasubramanian K (2012) Experimental investigation and flow visualization to determine the optimum dimension range of microgap heat sinks. Int J Heat Mass Transf 55:7623–7634

Chapter 8
Two-Phase Microgap Channel Cooling Technology for Hotspots Mitigation

Keywords Microgap channel · Microchannel · Flow boiling · Hotspots mitigation

Hotspots can be generated by non-uniform heat flux condition over the heated surface due to higher packaging densities and greater power consumption of high-performance computing technology in military systems designs. Because of this hotspot within a given chip, local heat generation rate exceed the average value on the chip and increase the peak temperature for a given total power generation which degrades the reliability and performance of equipments. This chapter presents the comparison of hotspot mitigation ability of microgap heat sink with conventional straight microchannel heat sink at the beginning of the chapter. The comparison is done with the same footprint, the same inlet mass flux, and the same effective and wall heat flux supplied based on the footprint. In the later section, this chapter presents the influencing factors of hotspots and its mitigation in microgap heat sink.

8.1 Comparison of Hotspot Mitigation Ability of Microgap Channel with Microchannel Heat Sink

Non-uniform heat flux conditions were applied to experimentally investigate the applicability of microgap heat sink for minimizing temperature gradient and mitigating hotspots from the heated surface of electronic device.

Temperature variations on chip wall in microchannel and microgap heat sink are shown in Fig. 8.1. In this figure, wall temperatures were taken at five spanwise locations with the diode sensors 5, 10, 15, 20, and 25 as shown in figure. Higher heat flux was maintained at the middle diode and heater, 15 to create a hotspot. Experimental results show that microgap maintains uniform wall temperature and minimize temperature gradient rather than microchannel at same mass flux near hotspot. According to Koo et al. [1] pressure drop is the most critical factor in

T. Alam et al., *Flow Boiling in Microgap Channels*,
SpringerBriefs in Thermal Engineering and Applied Science,
DOI: 10.1007/978-1-4614-7190-5_8, © The Author(s) 2014

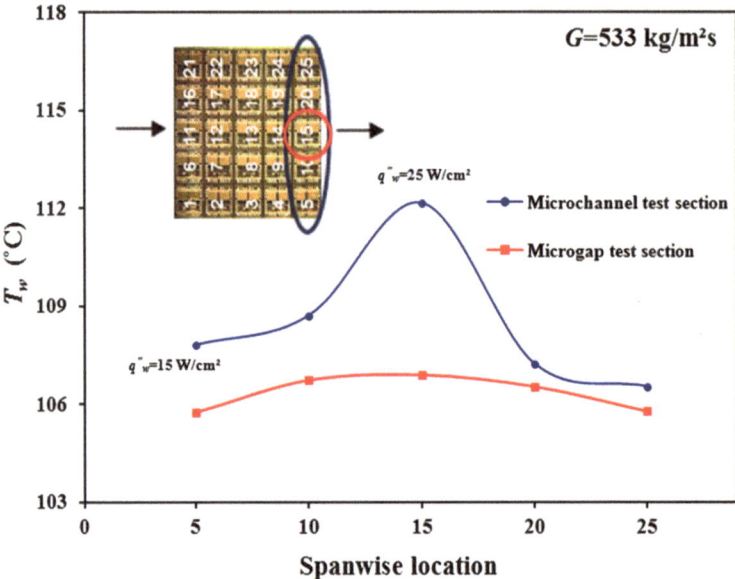

Fig. 8.1 Comparison of local wall temperature with hotspot between microchannel and microgap at $G = 533$ kg/m^{2}s

design of microchannel heat sink and they suggested that optimization should be performed to minimize the pressure drop along the microchannels to reduce temperature variations. In microgap heat sink, pressure drop is low and maintain a uniform pressure field through the surface whereas in microchannel, large pressure drop as shown in Fig. 4.6 may lead large variation of heated wall temperature.

8.2 Influencing Factors on Hotspots Mitigation in Microgap Heat Sink

The influence of mass fluxes on wall temperature with hotspot for microgap test section is shown in Fig. 8.2. Wall temperatures were taken at five spanwise locations with the diode sensors 5, 10, 15, 20, and 25 as shown in figure. Higher heat flux was maintained at the middle diode and heater, 15 to create a hotspot. From figure, it can be seen that the wall temperature near hotspot decrease with decrease of mass flux. This result can be explained as—the vapor quality at the outlet is higher for a lower mass flux at a given heat flux. Because of these increases of vapor quality, earlier transitions to annular flow for lower mass flux at a fixed heat flux occur. Annular flow, which is associated with thin liquid layers flowing along the outer walls of the channel and the vapor flows in the centre of the channel called vapor core. This is thermally advantageous, due to the high heat

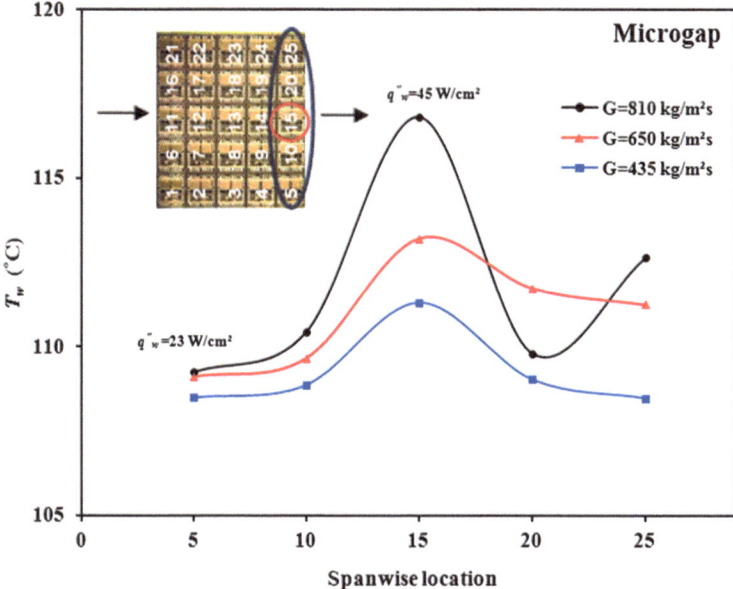

Fig. 8.2 Effect of mass flux on local wall temperature with hotspot at microgap heat sink

transfer rates associated with the evaporation of thin liquid layers. Thin liquid layers have low resistance to thermal diffusion and evaporation of liquid into the vapor core can promote the removal of substantial thermal energy from the walls. As the layer thins, the heat transfer rate increases and lower mass flux reduce the temperature gradient over the heated surface.

Minimization of temperature gradient and reduction of maximum temperature on the heated surface of the device are the two important objectives in electronic cooling. Figure 8.3 shows the test chip wall temperature variation for different depth of microgap at heat flux 28 W/cm² with a hotspot of heat flux 52 W/cm² and mass flux 435 kg/m²s. It can be seen from the figure that smaller the gap depth; more uniform the wall temperature and minimum the temperature gradient. Moreover, smaller gap maintain lower wall temperature. This is due to the smaller microgap size relative to the bubble diameter at departure; bubbles occupying the microgap create confinement effects. So, instead of nucleate boiling, evaporation of thin liquid layer removes more heat from wall and maintains lower wall temperature for smaller gap. Moreover, micro-layer thickness is strongly affected by the gap size, and decreases with decreasing gap size. Therefore, in the micro-layer dominant region, the vaporization rate is increased, and higher boiling heat transfer is possible due to the thinner micro-layer in smaller gap which interns maintain lower wall temperature in these gaps.

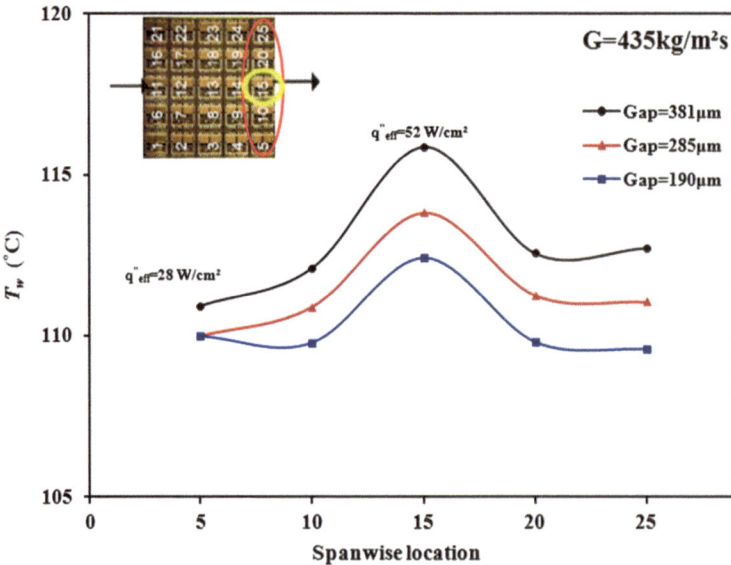

Fig. 8.3 Comparison of local wall temperature with hotspot at diode position 15 for different gap sizes at $G = 435$ kg/m^{2}s

Reference

1. Koo J, Jiang L, Bari A, Zhang L, Wang E, Kenny TW, Santiago JG, Goodson KE (2002) Convective boiling in microchannel heat sinks with spatially-varying heat generation. Thermal and Thermomechanical Phenomena in Electronic Systems (2002), ITherm, pp 341–346

Chapter 9
Conclusions and Recommendations

Key accomplishments from the present study, significances and limitations of this study, and recommendations for future work are summarized in this chapter.

9.1 Conclusions

Flow boiling experiments of deionized water in silicon microgap test sections were performed. Microgap heat sinks over a range of gap dimensions, mass fluxes, heat fluxes, and surface roughnesses were used to investigate the local flow boiling heat transfer and pressure drop characteristics in microgap channel and also to determine the most effective and efficient range of microgap dimensions and operating conditions based on heat transfer and pressure drop performance. The heat transfer, two-phase pressure drop, instabilities, and hotspots mitigation performances of the microgap channel were compared with the straight microchannel geometry. Simultaneous high speed flow visualizations were also conducted along with experiments to illustrate the bubble characteristics in the boiling flow in microgaps.

Key findings from these extensive microgap boiling experiments and flow visualizations are summarized as follows:

- Confinement in flow boiling occurs for microgap sizes 500 μm and below whereas physical confinement does not occur for microgaps 700 μm and above. For microgaps 700 μm and above, bubbly flow is the dominant flow regime at low heat fluxes and slug/annular flow regime forms as a result of bubbles coalescence as the heat flux is increased. For microgaps 500 μm and below, confined slug is the dominant flow regime at low heat fluxes and confined annular flow regime forms as the heat flux is increased.
- Smaller microgaps of dimension range 100–500 μm are very effective as they maintain very uniform and low wall temperature all over the heated surface before dryout takes places. However, below 100 μm sized microgaps are ineffective as partial dryout strikes very early. Heat transfer coefficient increases

T. Alam et al., *Flow Boiling in Microgap Channels*,
SpringerBriefs in Thermal Engineering and Applied Science,
DOI: 10.1007/978-1-4614-7190-5_9, © The Author(s) 2014

with the decreases of gap dimension and higher heat transfer coefficient is achieved within the range of microgaps 100–500 µm before dryout takes place. Heat transfer coefficient becomes independent of microgap dimension above 700 µm. In addition, pressure drop for microgap 200 µm and above is much smaller and increases slightly with the increase of heat flux.

- Microchannel heat sink gives better heat transfer performance at low heat flux due to early establishment of slug/annular flow. In contrast, microgap heat sink performs better at high heat flux due to confined slug and annular boiling dominance and consequent delay of dryout phase. Pressure drop increases gradually with heat flux for both microchannel and microgap heat sink. However, pressure drop is higher in microchannel than microgap heat sink at all the heat fluxes. Furthermore, microgap heat sink maintains uniform wall temperature and minimizes temperature gradient throughout the microgap wall at all heat and mass flux condition compare to the microchannel heat sink.

- Flow visualization reveals that surface roughness shows significant effect on boiling incipience and bubble nucleation site density increases with the increase of surface roughness from $R_a = 0.6$ to 1.0 µm for larger microgap heat sink. Microgaps of surface roughnesses, $R_a = 1.0$ and 1.6 µm show similar boiling behavior. However, surface roughness effect is less marked as dimension of microgap heat sink is reduced. Lower wall temperature is needed to commence boiling over the heated surface for rougher surface and a lower chip wall temperature is maintained at a fixed heat flux for both single and two-phase regions for microgap of surface roughness 1 µm and above. No significant adverse effect of surface roughness on pressure drop curve is observed at current tested microgap heat sink.

- Unlike microchannel, microgap allows vapor bubble to grow in both axial and radial direction instead of being forced upstream which minimize the pressure instability. Moreover, microgap maintains much lower wall temperature fluctuation than microchannel heat sink. A cycle includes bubble nucleation, expansion to both upstream and downstream, partial dryout and rewetting is observed in microgap. The duration of this cycle decreases as the microgap size decreases. Inlet pressure fluctuation increases with the decreasing microgap size due to the shape of vapor slug and also due to the decreasing nucleation cycle duration.

- The inlet pressure fluctuation increases as the heat flux increases and mass flux decreases. However, wall temperature fluctuation shows a mixed behavior on heat and mass flux. Moreover, surface roughness appears to have an adverse effect on the inlet pressure instability and $R_a = 1.0$ and 1.6 µm surfaces have between 20 and 40 % higher inlet pressure instability than $R_a = 0.6$ µm surface. However, wall temperature fluctuations are independent of surface roughness for microgap at various imposed heat fluxes and mass fluxes.

- Due to convective boiling nature in microgap, evaporation of thin liquid into the vapor core can promote the removal of substantial thermal energy from the walls and shows the potential of microgap heat sink for hotspots mitigation and lower mass flux as well as smaller depth microgap promote better hotspot mitigation performance.

9.2 Significances of This Study

The significance of the two-phase microgap cooling technique in view of its excellent performance in heat transfer, hotspots mitigation and instabilities reduction, is that this technique is suitable for thermal management in compact spaces with a smaller rate of coolant flow in high performance electronic devices. One implication is that thermal management of an electronic device like a defense application that calls for high capacity cooling is the High Energy Laser (HEL) system due to large power requirements and high heat fluxes (500 W/cm^2) are now possible. Compared to the other thermal management techniques such as microchannel based heat sink, this Si based microgap technique has the advantage that it eliminates the problems associated with interface thermal resistance. Microgap heat sink is easy to fabricate as it requires no external attachment to cut channel and micromachining; fluid can flow on the back surface of an active electronic component which in turn reduces interface thermal resistance. In addition, the unique thin film boiling mechanism in optimized microgap heat sink could provide a new pathway for hotspot and instability mitigation in high performance electronic devices.

9.3 Recommendations for Future Work

It should be noted that the experimental study presented here has some limitations. First, this study only takes account of the microgap dimensions from a range of 80–1,000 µm at a mass flux range of 400–1,000 kg/m^{2}s and maximum imposed effective heat flux ranging from 0 to 100 W/cm^2. The heat flux to the chip is increased from zero to the point at which the maximum wall temperature reaches 150 °C. As the solder bumps in the test chip may melt and damage the test chip above this temperature, wall temperature as well as heat flux are limited to this range. It should also be noted that only deionized water is used as a coolant in this microgap study; other coolants like FC-72, FC-77, and R-134a are not considered in this research.

Further research is needed to extend the microgap dimension range, mass flux and heat flux range using different types of coolants to understand the underline mechanisms behind this technique. An improved test chip conjugate with high temperature sustainable solder bump may help to extend these experimental conditions. Another interesting area for future work is the development of the two-phase flow regime based heat transfer and pressure drop model for microgap heat sink by correlating the experimental results with the high speed visual data.

Appendix A
Uncertainty Analysis for Experimental Data

To compute the uncertainty in the experimental data of this work, error analyses have been conducted according to the principles proposed by Taylor [1]. The error analysis procedures are summarized below:

Uncertainty in Sums and Differences

Suppose that $x,...,$ w are measured with uncertainties δx, ..., δw, and the measured values used to compute

$$f = x + \cdots + z - (u + \cdots + w)$$

If the uncertainties in x, ..., w are known to be independent and random, then the uncertainty in f is the quadratic sum of the original uncertainties.

$$\delta f = \sqrt{(\delta x)^2 + \cdots + (\delta z)^2 + (\delta u)^2 + \cdots + (\delta w)^2}$$

In any case, δf is never larger than their ordinary sum,

$$\delta f \leq \delta x + \cdots + \delta z + \delta u + \cdots + \delta w$$

Uncertainties in Products and Quotients

Suppose that $x,...,$ w are measured with uncertainties $\delta x,...,$ δw, and the measured values used to compute

$$f = \frac{x \times \cdots \times z}{u \times \cdots \times w}$$

If the uncertainties in x, ..., w are independent and random, then the fractional uncertainty in f is the sum in quadrature of the original fractional uncertainties,

T. Alam et al., *Flow Boiling in Microgap Channels*,
SpringerBriefs in Thermal Engineering and Applied Science,
DOI: 10.1007/978-1-4614-7190-5, © The Author(s) 2014

$$\frac{\delta f}{|f|} = \sqrt{\left(\frac{\delta x}{|x|}\right)^2 + \cdots + \left(\frac{\delta z}{|z|}\right)^2 + \left(\frac{\delta u}{|u|}\right)^2 + \cdots + \left(\frac{\delta w}{|w|}\right)^2}$$

In any case, it is never larger than their ordinary sum,

$$\frac{\delta f}{|f|} \leq \frac{\delta x}{|x|} + \cdots + \frac{\delta z}{|z|} + \frac{\delta u}{|u|} + \cdots + \frac{\delta w}{|w|}$$

Uncertainty in Any Function of One Variable

If x is measured with uncertainty δx and is used to calculate the function $f(x)$, then the uncertainty δf is

$$\delta f = \left|\frac{df}{dx}\right| \delta x$$

Uncertainty in a Power

If x is measured with uncertainty δx and is used to calculate the power $f = x^n$ (where n is a fixed, known number), then the fractional uncertainty in f is $|n|$ times that in x,

$$\frac{\delta f}{|f|} = |n| \frac{\delta x}{|x|}$$

Uncertainty in a Function of Several Variables

Suppose that $x,\ldots,$ z are measured with uncertainties $\delta x,\ldots,$ δz, and the measured values used to compute the function $f(x, \ldots, z)$. If the uncertainties in $x,\ldots, z$ are independent and random, then the uncertainty in f is

$$\delta f = \sqrt{\left(\frac{\partial f}{\partial x}\delta x\right)^2 + \cdots + \left(\frac{\partial f}{\partial z}\delta z\right)^2}$$

In any case, it is never larger than their ordinary sum,

Table A.1 The measurement accuracies and experimental uncertainties associated with sensors and parameters

Sensors and parameters	Accuracies and uncertainties
T-type thermocouples	±0.5 °C
Diode temperature sensors	±0.3 °C
Flow meter	±5 ml/min
Pressure transducer	±1.8 mbar
Differential pressure transducer	±1 mbar
Voltage measurement	±0.06 V
Current measurement	±0.15 A
Dimension measurement	±10 μm
Heat flux	2–8 %
Pressure drop	4–18 %
Heat transfer coefficient	4–10 %

$$\delta f \leq \left|\frac{\partial f}{\partial x}\right|\delta x + \cdots + \left|\frac{\partial f}{\partial z}\right|\delta z$$

Table A.1 shows the measurement accuracies and experimental uncertainties associated with sensors and parameters.

Table A.2 shows a set of uncertainty values in different parameters calculated based on the above equations for 300 μm depth microgap at mass flux, $G = 390$ kg/m^{2}s and heat flux, $q''_{eff} = 52.5$ W/cm^2.

Table A.2 Sample uncertainty calculation for 300 μm depth microgap at mass flux, $G = 390$ kg/m²s and heat flux, $q''_{eff} = 52.5$ W/cm²

Width (W)	Length (L)	Thickness (t)	Voltage (V)	Current (I)	T_f	T_d	K_s
1.27 ± 0.001 cm	1.27 ± 0.001 cm	0.0675 ± 0.001 cm	10.7 ± 0.06 V	8.5 ± 0.15 A	101.76 ± 0.5 °C	119.6 °C ± 0.3 °C	1.21 W/cm °C

Calculation

$$q''_{eff} = \frac{q_{eff}}{A} = f(V, I, W, L) = 52.5 \text{ W/cm}^2$$

$$\frac{\delta q''_{eff}}{\left|q''_{eff}\right|} = \sqrt{\left(\frac{\delta V}{|V|}\right)^2 + \left(\frac{\delta I}{|I|}\right)^2 + \left(\frac{\delta W}{|W|}\right)^2 + \left(\frac{\delta L}{|L|}\right)^2} = 0.01855 = 1.855\% \approx 2\%$$

$$T_f = T_{sat}$$

$$\delta T_f = \delta T_{sat} = \pm 0.5 °C$$

$$T_w = T_d - \frac{q''_{eff} t}{K_s} = 116.77 °C$$

$$\delta T_w = \sqrt{\left(\frac{\partial T_w}{\partial T_d} . \delta T_d\right)^2 + \left(\frac{\partial T_w}{\partial q''_{eff}} . \delta q''_{eff}\right)^2 + \left(\frac{\partial T_w}{\partial t} . \delta t\right)^2 + \left(\frac{\partial T_w}{\partial K_s} . \delta K_s\right)^2}$$

$$= \sqrt{(1.\delta T_d)^2 + \left(-\frac{t}{K_s} . \delta q''_{eff}\right)^2 + \left(-\frac{q''_{eff}}{K_s} . \delta t\right)^2 + 0} = \pm 0.3 °C$$

$$\Delta T = T_w - T_f = 15.005 °C$$

$$\delta(T_w - T_f) = \delta \Delta T = \sqrt{(\delta T_w)^2 + (\delta T_f)^2} = \pm 0.6 °C$$

$$h_z = \frac{q''_{eff}}{\Delta T}$$

$$\frac{\delta h_z}{|h_z|} = \sqrt{\left(\frac{\delta q''_{eff}}{q''_{eff}}\right)^2 + \left(\frac{\delta \Delta T}{|\Delta T|}\right)^2} = 0.038 = 3.8\% \approx 4\%$$

Appendix B
Nomenclature

A	Footprint area (cm^2)
A_c	Wetted area of microchannel (cm^2)
A_{gap}	Microgap cross-sectional area (cm^2)
A_{man}	Manifold cross-sectional area (cm^2)
Bl	Boiling number
Bo	Bond number
Co	Confinement number
c_p	Specific heat, (J/kg °C)
d	Depth of microchannel (μm)
D	Microgap depth (μm)
g	Gravitational acceleration
G	Mass flux (kg/m^2s)
h	Heat transfer coefficient (W/m^2K)
h_{fg}	Heat of vaporization (J/kg)
k_s	Thermal conductivity, W/cm °C
K_c	Loss coefficient
L	Length of the substrate (cm)
$\dot{m}$	Mass flow rate (kg/s)
N	Number of microchannels
P	Pressure (bar)
ΔP	Pressure drop (bar)
q	Total heat dissipation (W)
q_{eff}	Effective heat dissipation (W)
q''_{eff}	Effective heat flux (W/cm^2)
q_{loss}	Heat loss (W)
R_a	Roughness parameter (arithmetic mean value)
Re	Reynolds number
R_t	Roughness parameter (maximum peak to valley height)
t	Substrate thickness (cm)
T	Temperature (°C)

T. Alam et al., *Flow Boiling in Microgap Channels*,
SpringerBriefs in Thermal Engineering and Applied Science,
DOI: 10.1007/978-1-4614-7190-5, © The Author(s) 2014

V_d Voltage drop across diode (V)
w Channel width (μm)
W Width of the substrate (cm)
x Vapor quality
z z-Coordinate (axial distance) (cm)

Greek Symbols

ρ Density (kg/m^3)
μ Dynamic viscosity (Ns/m^2)
σ Surface tension (N/m)
η Fin efficiency

Subscripts

c   Contraction
d   Diode
e   Expansion
f   Liquid
g   Vapor
gap   Microgap
i   Manifold inlet
o   Manifold outlet
man   Manifold
s   Substrate
sat   Saturated
sp   Single-phase
w   Wall
z   Local

Appendix C
Data Reduction

Heat Transfer Data Reduction

The effective heat supplied, q_{eff} to the fluid in each test piece and the effective heat flux q''_{eff} is calculated as given.

The effective heat transfer rate, q_{eff} to the fluid in microgap channel is obtained by:

$$q_{eff} = q - q_{loss} \qquad (C.1)$$

Where q is input power and q_{loss} is heat loss.

The effective heat flux q''_{eff} that the heat sink can dissipate is calculated from:

$$q''_{eff} = \frac{q_{eff}}{A} \qquad (C.2)$$

where A is the base area of heat sink, $A = W \times L$.

For microchannel, the total wetted area of the microchannels is:

$$A_c = N(w + 2\eta d)L \qquad (C.3)$$

where N is total number of channels; w, d and L are the width, depth, and length of the channel respectively and η is the efficiency of a fin with adiabatic tip which is correlated by:

$$\eta = \frac{\tanh(md)}{md} \qquad (C.4)$$

and

$$m = \sqrt{\frac{2h}{K_s w_w}} \qquad (C.5)$$

where K_s is the thermal conductivity of the substrate and w_w is the width of the channel wall.

So, the wall heat flux for microchannel is defined as

$$q''_w = \frac{q_{eff}}{A_c} \qquad (C.6)$$

T. Alam et al., *Flow Boiling in Microgap Channels*,
SpringerBriefs in Thermal Engineering and Applied Science,
DOI: 10.1007/978-1-4614-7190-5, © The Author(s) 2014

The test section is divided into two regions: an upstream subcooled inlet region and a downstream saturated region as subcooled ($T_{f,i} < T_{sat}$) water is supplied into the heat sink for all test conditions; the location of zero thermodynamic equilibrium quality ($x = 0$) serves as a dividing point between the two regions.

The local heat transfer coefficient in microgap is calculated from,

$$h_z = \frac{q''_{eff}}{T_w - T_f} \tag{C.7}$$

The local heat transfer coefficient in microchannel is calculated from,

$$h_z = \frac{q_{eff}}{A_c(T_w - T_f)} \tag{C.8}$$

in which T_f is the fluid temperature as defined by

$$T_f = T_{f,i} + \frac{q''_{eff}Wz}{\dot{m}c_p} \quad (\text{single} - \text{phase region}) \tag{C.9}$$

where z, $\dot{m}$ and c_p are the axial distance, mass flow rate, and specific heat respectively.

$$T_f = T_{sat}(\text{saturated region}) \tag{C.10}$$

T_w, is the local wall temperature. This temperature is corrected assuming one dimensional heat conduction through the substrate

$$T_w = T_d - \frac{q''_{eff}t}{K_s} (\text{for microgap}) \tag{C.11}$$

$$T_w = T_d - \frac{q''_{eff}(t - d)}{K_s} (\text{for microchannel}) \tag{C.12}$$

where t and K_s are the substrate thickness and thermal conductivity respectively. T_d is the measured temperature by an integrated diode.

Bond number is defined as the ratio of buoyancy force to surface tension force.

$$Bo = \left[\frac{g(\rho_f - \rho_g)}{\sigma}\right]D^2 \tag{C.13}$$

where σ is the surface tension, g is the gravitational acceleration, ρ_f and ρ_g are liquid and vapor densities of fluid respectively. D is the gap depth. Some other non-dimensional parameter like Boiling number, Bl which is non-dimensional heat flux and Reynolds number, Re are defined as follows:

$$Bl = \frac{q''_{eff}}{Gh_{fg}} \tag{C.14}$$

$$Re = \frac{GD}{\mu_f} \tag{C.15}$$

where h_{fg} and μ_f are the heat of vaporization and dynamic viscosity of fluid respectively.

Pressure Drop Data Reduction

Pressure taps are located across the microgap and microchannel inlet and outlet plenum. These taps are positioned as close as possible to the test die. Pressure losses by the sudden contraction (ΔP_c) and the sudden enlargement (ΔP_e) were very small compared with the frictional pressure drop. Though these values are very small of total pressure changes, the pressure drop and the pressure recovery at the sudden contraction and the sudden enlargement were considered for calculation of the total pressure drop.

Pressure losses are calculated based on the methods described in Blevins [2], Chislom and Sutherland [3] and Collier and Thome [4]. As mentioned earlier, subcooled water ($T_{f,\ i} < T_{sat}$) is supplied into the heat sink for all test conditions. The pressure drop associated with the liquid flow at the sudden contraction in microgap channel is calculated as

$$\Delta P_c = \frac{G^2}{2\rho_f}\left[1 - \left(\frac{A_{gap}}{A_{man}}\right)^2 + K_c\right] \tag{C.16}$$

where G is mass flux in the microgap, ρ_f is liquid density and K_c is the non-recoverable loss coefficient for laminar flow given by

$$K_c = 19\left(\frac{\mu_f}{GD}\right) + 0.64 \tag{C.17}$$

The pressure recovery at the sudden enlargement at the exit is calculated as

$$\Delta P_e = \frac{G^2}{\rho_f}\left(\frac{A_{gap}}{A_{man}}\right)\left[1 - \left(\frac{A_{gap}}{A_{man}}\right)\right]\left[1 + \left(\frac{\rho_f}{\rho_g} - 1\right)x\right] \tag{C.18}$$

The microchannel pressure drops (ΔP) are calculated as follows. The pressure drop associated with the liquid flow at the sudden contraction is calculated as

$$\Delta P_c = \frac{G^2}{2\rho_f}\left[1 - \left(\frac{NA_{ch}}{A_{man}}\right)^2 + K_c\right] \tag{C.19}$$

where G is mass flux in the microgap, ρ_f is liquid density and K_c is the non-recoverable loss coefficient for laminar flow given by

$$K_c = 0.0088\left(\frac{d}{w}\right)^2 - 0.1785\left(\frac{d}{w}\right) + 1.6027 \tag{C.20}$$

The pressure recovery at the sudden enlargement at the exit is calculated as

$$\Delta P_e = \frac{G^2}{\rho_f}\left(\frac{NA_{ch}}{A_{man}}\right)\left[1 - \left(\frac{NA_{ch}}{A_{man}}\right)\right]\left[1 + \left(\frac{\rho_f}{\rho_g} - 1\right)x\right] \qquad (C.21)$$

Therefore, the pressure drops (ΔP) reported below are

$$\Delta P = [(P_i - \Delta P_c) - (P_o + \Delta P_e)] \qquad (C.22)$$

References

1. Taylor JR (1997) An introduction to error analysis: The study of uncertainties in physical measurements, University Science Books, 2nd edn. US
2. Blevins RD (1991) Applied fluid dynamics handbook. Krieger Publishing Co., Berlin, pp 77–78
3. Chislom D, Sutherland LA (1969) Prediction of pressure gradients in pipeline systems during two-phase flow. Symposium in two-phase flow systems. University of Leeds
4. Collier JG, Thome JR (1994) Convective boiling and condensation. Clarendon Press, Oxford